QUANTUM PARTICLES OF GRAVITY

Noah MacKay

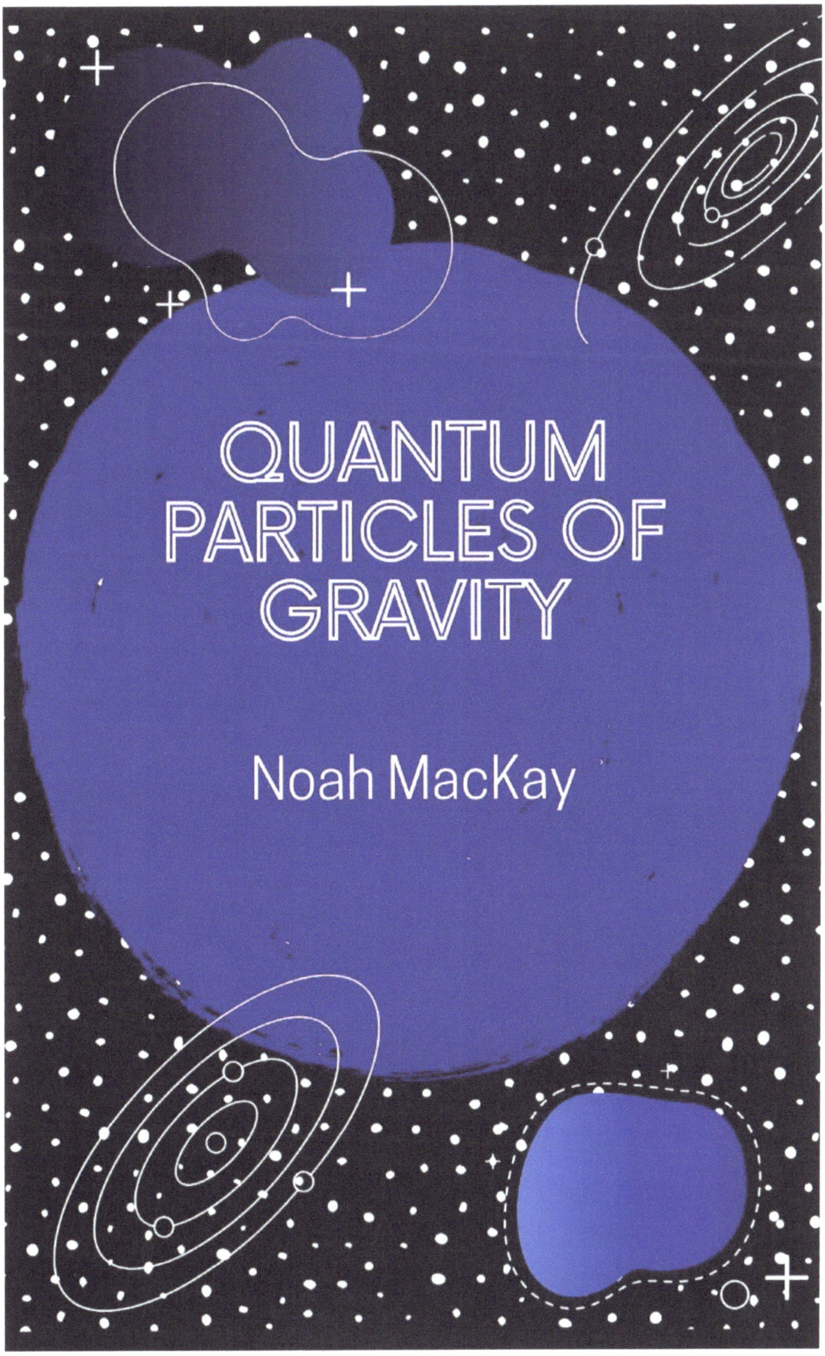

Quantum Particles of Gravity

by Noah MacKay

2021

TABLE OF CONTENTS

QUANTUM PARTICLES OF GRAVITY

QUANTUM PARTICLES OF GRAVITY

NOTES FROM THE AUTHOR

It was Richard Feynman who once said, "If you cannot describe a concept of physics in layman's terms, you do not know the concept yourself." As I am writing this, and thinking about how to progress with this book, the more Feynman is proven to be right. The subject of quantum gravity is a realm of mystery and uncertainty; it is honestly difficult to write a book on such tentative topics, only to eventually realize their viewpoints just might have been disproven by science. Over the past half a century, the theories of string theory and loop quantum gravity were drafted to connect quantum mechanics and Albert Einstein's general theory of relativity. As spot on they might be on their own, the wider the gap between these two theories becomes. The question I am concerned with in quantum gravity is, basically, whether or not a bridge can be built to cross the chasm.

I recall back in the spring of 2017; I was a freshman at East Carolina University studying physics and German. One night I sat on a bench in front of my dormitory and I stared up at the full moon. Spring nights in eastern North Carolina, especially approaching summer, are usually humid. But that particular night was gorgeous. I stared at the moon and the surrounding stars, and my mind went adrift in deep thought. In the fall of 2016 I heard of the groundbreaking news of gravitational wave detection from the LIGO observatories. Around late spring I had just learned of a fundamental concept of quantum physics called "wave-particle duality:" particles share wave mechanics, and waves obtain particle attributes. As I was in deep thought, I asked myself, *what if GRAVITY had a WAVE-PARTICLE DUALITY?*

This hypothetical *Quantum Particle of Gravity* must be the keystone linking string theory and loop quantum gravity, I thought (and still thinking now). And perhaps future research may very well verify their detection and unravel how they behave in the universe and our lives. This book is intended to discuss the nature and properties of the *GRAVITON*: the quantum particle of gravity. In addition, I will also discuss its role in the conventional theories of string theory and in loop quantum gravity. Being that I am writing for the public reader, I will offer very little mathematics than I would like to, as I try to be concise in my scientific writing. However, be ready to see equations that may look intimidating. Just know, what the math is saying in numerals and Greek, I am translating them to English.

As an author writing to the general public, I do not expect you to know gravitons, string theory or loop quantum gravity beforehand. What I do expect you to know is what gravity is in the general sense, along with some basic idea of what quantum mechanics is about. If these prerequisites are not met, then I suggest you read the following books before this one:

The Little Book of Black Holes by Steven Gubser

A Brief History of Time by Stephen Hawking

Quantum Mechanics: The Theoretical Minimum by Leonard Susskind

Special Relativity and Classical Field Theory by Leonard Susskind

The Theoretical Minimum by Leonard Susskind

The Road to Reality by Roger Penrose

In addition I recommend these books for the subject of quantum gravity from other writers:

The Little Book of String Theory by Steven Gubser

Reality Is Not What It Seems by Carlo Rovelli

What you would expect in this book is an introduction to gravity as a classical theory, beginning with Newton's theory and its contributing breakthroughs. I will then discuss Einstein's theory of gravity; how general relativity rectified the problems with Newton's theory and how it led to the groundbreaking findings of Schwarzschild and Hawking.

Afterwards I will discuss the basis of quantum mechanics for all particles: the notion of energy quanta, wave-particle duality, and the roles of the "Hamiltonian" and "total angular momentum." This is where I will introduce gravity as a quantum theory, discussing the motivations and goals of string theory and loop quantum gravity (while applying how the graviton is treated in these theories). I will mention the successes and extensions to each theory, as well as their limitations and where their merits end. Then finally I will introduce the graviton: the titular character of this book. Along with the two main theories of quantum gravity, I will offer a detailed history and description of the graviton, its nature, and its properties. I will discuss all possible theories of the graviton, itself, leading up to the present-day tentative research involving gravitons. To end the book, I will offer a layman's course in gravitational wave-particle duality, how the graviton might be detectable in gravitational waves.

INTRODUCTION:
UNDERSTANDING NATURE

Physics is one of the fundamental sciences of our universe. The universe we know of is dictated by the laws of science, explicitly biology, chemistry, and physics. Biology answers the question: *What is that thing?* Chemistry answers the question: *What is that thing made of?* And finally physics answers the question: *What is that thing doing?*

This book covers the history, evolution, and the future of the theory of gravitation and our understanding thereof. A dominant force in the cosmos, from planets to stars, to galaxies and nebulae; gravitation orchestrates the behavior of the universe just as pure as Beethoven, Mozart, even Schubert orchestrates the symphony. A simple glance up at the night sky, to see the heavens unfold with the glisten of the stars and the shine of the pale full moon, one must wonder how everything works so beautifully and divinely - as I have once wondered.

Across human history, it has been the goal for humanity to answer the ultimate questions of existence: Why are we here? Where do we come from? It is within human nature to devise a meaningful theory or a worldview of nature around us, to find some sense of uniformity in a world full of chaos and disorder.

The ancient societies of Europe and India turned to a religious worldview. The ancients believed that the phenomena of nature were divine interventions manifested by one almighty God of Judeo-Christian monotheism, or by one of the many gods of Indo-European polytheism.

Consider the archaic art epochs of Ancient Greece. The Cycladic bust captured the basic shape of the human face, however without the ridges of the nose or the depth of the eyelids.

The Cycladic Bust

Once the rest of the human body was better understood, the kouros/kore sculptures displayed the finer details of edging, muscle tone and the braided hair. However, even the kouros sculptures were bland, emotionless, and stiff. And finally, once physique and emotion were mastered, the sculpting was perfected in the era of classical antiquity.

The kouros (upper half) sculpture (left) and the Venus of Milo of the Greek classic antiquity (right).

The evolution of scientific understanding, and even the understanding of the theory of gravity, follows that same path. Each time frame of scientific understanding and technology, much like the archaic art epochs, presented a theory or worldview that best represented the universe at the time.

To make better sense of the world, the mythos of religion was approached with a scientific point of view, thus forming the logic-based logos rhetoric. The mathematical descriptions of Pythagoras, Archimedes, and Euclid - and the philosophical arguments of Anaximander, Heraclites and Aristotle - paved the way to our understanding on the nature of things we now take for granted.

As time and technology evolve, so must our understanding of nature. That is how the Earth, long believed to be flat, received its spherical shape. That is also how we learned that the universe, believed to be 6,000 years old, is actually 13.8 billion years old. But as we progressively move forward into the future, it is rather encouraged to look back at the past without complete regression.

If it were not the case, then the Renaissance would not have happened. The philosophies from ancient thinkers would resurface and thus inspire the scientific revolution, pioneered by Galileo, Kepler and Copernicus. This would in turn inspire Sir Isaac Newton, James Clarke Maxwell,

Albert Einstein, Stephen Hawking, and many other scientists on their scientific revolutions.

I want to make it clear that physics isn't a dead science, but a lively one whose goal is to solve this jigsaw puzzle that is our understanding of the universe.

Before Socrates, there was the Ancient Greek philosopher Anaximander, who is arguably called the first scientist. Although a philosopher, Anaximander approached the Ancient Greek mythos from a scientific point of view, which would form the basis of the logos rhetoric later carried forth by Pythagoras and Aristotle. It would be through this rhetoric that would inspire the practitioners of the Scientific Revolution to redefine the Catholic mythos into scientific fact. This had, in turn, led to the foundations of many widely accepted scientific theories, such as Galileo's principle of inertia, Newton's laws of motion and gravity, and Copernicus' heliocentric model.

The Pre-Socratic philosopher Anaximander.

Anaximander's Seven Bases of Nature, which was written in prose titled Περι φυσεως (*Peri physeós,* or *On Nature*), would outline how the philosopher understood nature. The Seven Bases are given as follows:

1. The occurrence of phenomena, and the transformation of one thing into another, are regulated by "necessity."

The first basis claims that physical phenomena and evolution influenced by time must be regulated by this so-called "necessity," or some fundamental law. This postulate was the first to acknowledge that nature is dictated by various laws of science.

2. The finite constituents of nature derive from an originating singularity, or the "apeiron," or "infinity".

This "apeiron" was the origin of all life. According to Anaximander, all substances in our experience can be understood in terms of something that is natural but, at the same time, is not one of the substances in our everyday life. This then considers the realistic question: "What is perception?" Are these substances of experience natural or supernatural?

To Anaximander, it must be both, supposing that what we see as a part of nature must have come from something supernatural, as if it were from divine origins. This is, of course, based on mythology that natural elements were given to humans by the Titans and Olympian Gods. I would assume that Anaximander would have taken a pantheistic (Nature is God) stance towards this claim of perception. Nonetheless, the "apeiron" could be considered as the first ancestor of the Big Bang theory, or even the Genesis creation myth, that the numerous finites originated from the singular infinite. It would be considered that the atom, the

Faraday force fields and other theoretical entities (i.e. quarks, General Relativity, Quantum Theory, etc.) all came from the same historic ancestor: the "apeiron."

The big bang, the modern example of the apeiron.

3. The world came to pass by the separation of the hot and cold forces from the "apeiron." The hot forces would then be divided to form the Sun, the Moon and the stars, whereas the cold would form the water that once covered the Earth.

4. The Earth is a multidimensional entity floating in space, suspended by the void, dominated by no other body.

5. The Sun, the Moon and the stars rotate around the Earth along hollow circular tracks, or wheels. What we would see as the Sun, the Moon and the stars is the exposed fire within these wheels. The Star wheel is closest to the Earth, the Moon wheel is in the middle, and the Sun wheel is farthest away, in a distance ratio of 9 : 18 : 27 (meaning at face value "very far," "still very far," "exceedingly far").

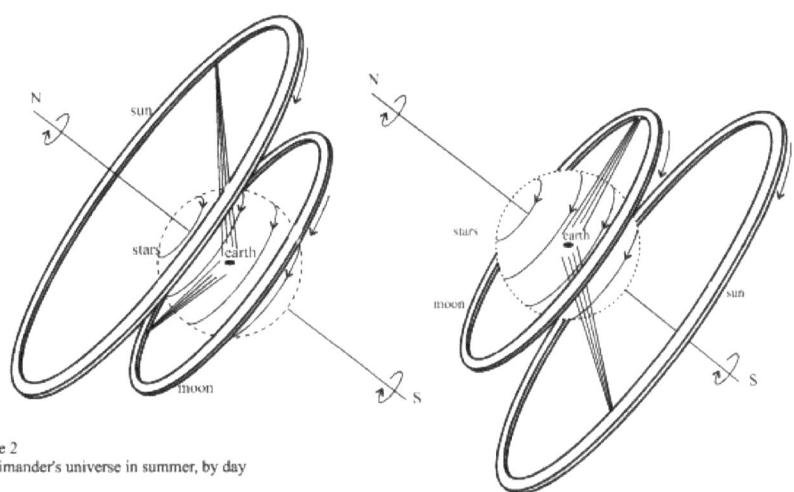

Figure 2
Anaximander's universe in summer, by day

Figure 3
Anaximander's universe in winter, by night

Anaximander's cosmology

The corresponding Ancient Greek myths include the formation of the world out of Chaos and the story of Atlas holding the world upon his shoulders. However, having these three bases as a description of the Earth and other celestial bodies was proof of a critical point regarding scientific progress: "The adventure of science is grounded by accumulating knowledge, but its soul is perpetual change."

The very essence of scientific thought is the avoidance of clinging onto worldviews flawed to modern standards; and instead changing them in light of knowledge, observation, discussion, different ideas, and criticisms. For Anaximander to reinterpret the creation myth and Ancient Greek cosmology with a scientific context with his "apeiron" and the fire wheels, these were revolutionary claims that made sense back then. But to today's modern standards it was flawed, for we know of orbital motion and the heliocentric model.

6. Meteorological phenomena have natural causes.

Anaximander was the first to recognize that rainwater was the water evaporated from the sea and the rivers, and that the earthquakes were fissures of the Earth (although claiming because of excessive heat or rain, with no prior thought of tectonic plate interactions). He also hypothesized that thunder and lightning were caused by the clashing and splitting of clouds.

7. All animals, including human beings, originally came from the sea that once covered the Earth, as descendants of fish or fishlike creatures.

Because Greece is practically surrounded by water, fish must have played an essential role to Ancient Greek life and the survival of its society. Because of its importance, this would logically explain why Anaximander would have conceived such a theory, that all creatures of nature (even humans) came from the sea. This may have been a coincidence, especially knowing thanks to biology that the first life came from microscopic organisms living in the oceans.

The scientific philosophy of Anaximander, as revolutionary as they were in his time, however lacked experimental observation and measurements – much like today's quantum theories of gravity. This would therefore discredit Anaximander as a scientist, or wouldn't it? This is the misconception of science, that quantitative results, experimental observation and mathematical techniques solidify a logical, scientific claim.

However, they are merely tools of a scientific theory, not the theory itself. The explicit goal of science is NOT to make correct quantitative predictions, but rather to simply understand how the world works (provided that the theory in question is rational, logical and refutable).

It is the perpetual process of continuous modification and improvement of a conceptual meaning of nature that outlines scientific advancement. A scientific theory would, therefore, have to have a reasonable philosophical hypothesis, which must have the potential to be bettered either by alteration or by disproof. On that note, I will introduce gravitation as a classical theory.

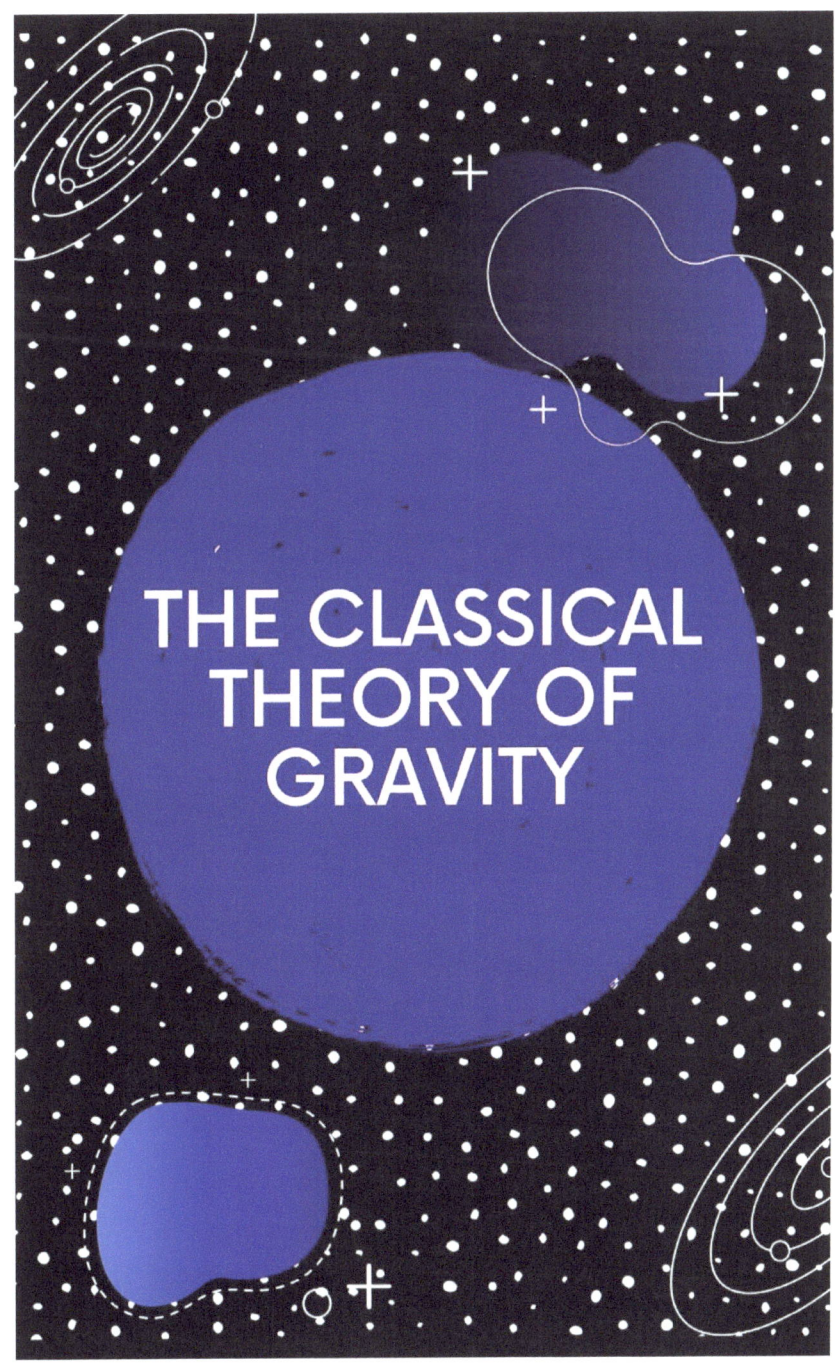

THE CLASSICAL THEORY OF GRAVITY

NEWTONIAN GRAVITY

In the ancient times, the strange phenomenon of gravitation has been a philosophical question notoriously in Greece and India. In Ancient Greece, the philosopher Archimedes hypothesized that the center of mass (the center of gravity) of any shape and object is in its geometric center. The ancient Indian philosophers Aryabhata and Brahmagupta both thought of gravity as the attractive force that is keeping us from being thrown outward by Earth's rotation; they called it "gurutvaakarshan."

In Ancient Rome, the architect and engineer Vitruvius hypothesized that the gravity of an object did not depend on its weight, but its "nature." These philosophical thoughts would eventually pave the way to the classical description of gravity, finally ending with the anecdote of the apple falling on Newton's head.

Sir Isaac Newton

Sir Isaac Newton was a Cambridge student by the time he formulated calculus and his three laws of motion. But once asked if his three laws could be applied in the universal scale, it took one apple to land on his head (or land beside him, most people go with the latter) to realize a great thought: free-falling and gravitational attraction is the same thing.

Before Newton was the Renaissance scientist Galileo Galilei, who was able to calculate the rate of free fall to be a constant acceleration of $g = 9.81 \text{ m/s}^2$.

Galileo Galilei

Before looking at "universal gravity," Newton only saw the force of gravity as the force of free-fall: $F = mg$ (the force of a falling object is the product between the object's mass and the rate of free-fall). But Newton's view on gravity must not only obey his three laws of motion, it would also expand upon his force of free-fall.

Newton's first law states that the law of inertia is applied to all objects: an object at rest remains at rest, and an object in motion remains in motion. Here on Earth this applies to resting, stationary objects as well as objects in uniform motion.

In outer space, a free-floating object could either be stationary or in uniform motion through the vacuum. Consider an asteroid; we cannot tell whether or not an asteroid is moving unless there are stationary stars to which we can compare its motions. If an asteroid is in deep space, in the black emptiness of the vacuum, there is no certainty whether or not that asteroid is moving at all.

The second law of motion is the law of force: to change inertial reference frames, there requires an external non-inertial force. Isaac Newton drafted this law with the accompanying equation: $F = ma$, where a is the acceleration of an object with mass m. Here on Earth, acceleration may mean thrust, tension, friction, and free-falling. Out in space, Newton's force of gravity is the mutual attraction between two objects. Because it is a force of attraction, gravity must be stronger when the two objects are drawn closer to each other. This expands the free-falling force as the "gravitational force,"

$$F_g = G \frac{m_1 m_2}{|r|^2}.$$

In every interaction involving universal gravity, two very large masses m_1 and m_2 are spaced apart by a distance $|r|$. But since objects can leave a planet's gravitational influence, the overall interaction is generally weak – however strengthened by an exceedingly large planetary mass. This general weakness of gravity comes from the gravitational constant $G = 6.67 \times 10^{-11}$ N $*$ m^2/kg^2.

This universal constant G, conventionally called the Newton constant, is also known as the Cavendish constant. Its value was first computed by English scientist Henry Cavendish in 1798 as a result of an experiment he performed. He observed the interactions between two small masses, attached to the ends of a thin suspended dumbbell, and two much larger adjacent masses. His experiment was a simulation of the gravitational interaction between astronomical objects. Because of the weakness of gravity between the objects, the constant resulted to a small number. Interesting enough, Newton was unaware of this constant's value. But he knew it connected his proportionalities into his law, and thus included it in his equations.

In a simple free-fall experiment, the test object is falling from a height that is a small deviation from the Earth radius. Putting this in Newton's gravitation law, the distance $|r|$ is the radius of the Earth $|R|$. Then, Newton's universal gravitational force becomes the force of free-fall:

$$m_1 g = G \frac{m_1 m_2}{|R|^2}.$$

Through Newton's force of gravity, this defines Galileo's rate of free-fall as the "gravitational field" of any celestial object with a large mass M:

$$g = G \frac{M}{|R|^2}.$$

Note that g is applicable for all planets and stars, as well as any massive object in the universe. For the Earth, the free-fall constant of $g = 9.81$ m/s^2 is also the constant rate of gravity on Earth. But this value is not the same for all massive objects, for every planet has a different mass and a different volume. This brings rise to the idea that larger masses have stronger gravities. For Jupiter of mass 1.90×10^{27} kg, its gravitational rate is 24.79 m/s^2, indeed stronger than the Earth's gravity.

Of course, for objects to leave a planet's gravitational pull, for instance NASA rockets, they have to move at such a speed that can overpower the gravitational rate. This is called the "escape velocity:"

$$v_{\text{esc}} = \sqrt{\frac{2GM}{|R|}},$$

which comes from the conservation of energy between a rocket's initial gravitational potential energy on the Earth's surface and its final kinetic energy in space.

And finally, Newton's third law of motion is the law of *accio-reaccio*, of action and reaction: the force of one reference object applied on another is equal but opposite to the force of the other object applied to the reference object. To Newton's law of gravity, the gravitational force between the Earth and the Moon is equal to the force between the Moon and the Earth:

$$m_1 G \frac{m_2}{|r|^2} = m_2 G \frac{m_1}{|r'|^2}.$$

This is only true when the two gravitational fields g_2 and g_1 are equal in strength. This is only possible at a singular point within a far distance between the two planets; a point at which the long distance is broken up between two distances of $|r|$ and $|r'|$. This point is called the "Lagrangian point."

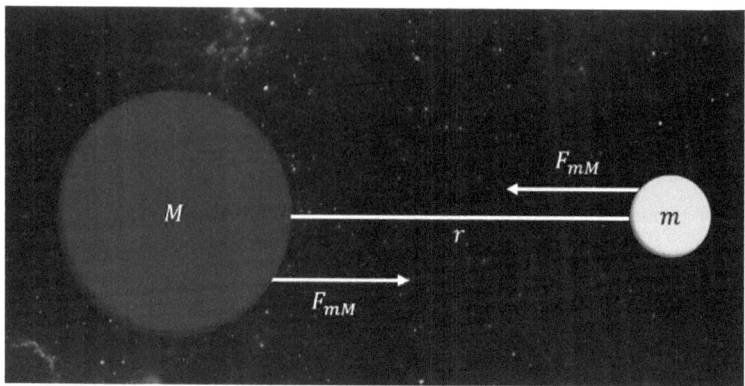

An example of universal gravitation.

As we saw, Newton's law of universal gravitation obeys his previous three laws of motion. And because of its accuracy and flawless mathematics, Isaac Newton was credited in discovering gravity as a physical force. Newton's law of gravity had inspired the derivation of more mathematics and had led to successful discoveries of additional planets.

German mathematician Johann Carl Friedrich Gauss derived the notion of "gravitational flux:" how much gravitational field is penetrating the surface area of a celestial object:

$$\Phi_g = 4\pi GM,$$

which increases linearly in proportion to the increase of mass.

Johann Carl Friedrich Gauss

This provides us an idea of the strength of gravity in relation to mass, alone. If normal, every-day objects we observe and encounter with here on Earth were sent to space, the gravitational flux they would induce would be significantly weaker than the gravitational flux of the Earth, of Jupiter, or even that of the Sun. This is why astronauts free-floating in space do not have their own gravitational fields like planets and stars do.

But the question now would be, *how much mass would it take to create gravitational fields*? This was a question Albert Einstein answered with his general theory

35

of relativity: his own theory of gravity (which will be covered soon).

Isaac Newton described gravity as a force, an interaction between a test mass and a source mass. But behind his mathematics and his theory of gravity as a force, Newton never said what gravity actually *was*. For instance, the force of a spring is the elastic stretching and compressing of a spring, which is a coiled wire of flexible and malleable metal. But the force of gravity is the mutual attraction of celestial objects with induced gravitational fields, which are (`insert definition of a gravitational field`).

That was deliberate.

The irony is, however, that the scientists of Newton's age until Einstein would care less of what gravity is, only how it affects the universe. Through Newton's math, scientific achievements would spawn in astronomy through the discoveries of Uranus in 1781 and Neptune in 1846. But there was one hiccup in Newton's theory of gravity: Mercury's eccentric orbit.

EINSTEINIAN GRAVITY

Albert Einstein rose to academic fame from the Swiss patent office in 1905, after he published four consecutive papers that kickstarted the movement of modern physics. His fourth and last 1905 paper was on the "electrodynamics of moving bodies," meaning the physics of objects moving close to the speed of light.

James Clerk Maxwell

Before Einstein was the Scottish physicist James Clerk Maxwell. Maxwell stated that the electric and magnetic fields, respectively E and B (previously thought to be separate and independent from one another), can be unified into an "electromagnetic field." The set of mathematical equations that would support this claim was self-derived, called the "Maxwell Equations:"

$$\nabla \cdot E = \frac{\rho}{\varepsilon_0}$$

$$\nabla \cdot B = 0$$

$$\nabla \times E = -\partial_t B$$

$$\nabla \times B = \mu_0 \varepsilon_0 \partial_t E + \mu_0 j$$

They may seem to be a jumbled mess of weird symbols and Greek letters, but these equations describe the flux and rotations of the electric and magnetic fields – how they are in some way related to one another. As we look at these equations, they ultimately depend on the environment of the interaction. An assortment of metals has a variety in electric and magnetic conductivity (copper is more conductive than lead, for example), just as how breathable air and the vacuum of outer space are just as different. Being that we were looking at gravity in outer space, let us look at the vacuum environment.

This is important because in the vacuum there is no charge or current densities, just the dynamic electric and magnetic fields:

$$\nabla \cdot E = 0$$
$$\nabla \cdot B = 0$$
$$\nabla \times E = -\partial_t B$$
$$\nabla \times B = \mu_0 \varepsilon_0 \partial_t E$$

The equations above are essential in describing these unified "electromagnetic fields." Maxwell even thought that the electric and magnetic fields can act as waves. Through these revised equations, James Clerk Maxwell derived these wave equations for the two fields:

$$\nabla^2 E = \mu_0 \varepsilon_0 \partial_t^2 E$$
$$\nabla^2 B = \mu_0 \varepsilon_0 \partial_t^2 B$$

which both share the fundamental constants of electricity and magnetism: ε_0 and μ_0.

Together, the fundamental constants of electricity and magnetism would combine and form an electromagnetic constant c. However, this coupled constant acts like speed, for waves are disbursed energy that propagate at some speed. This constant c is therefore a form of speed: the speed of light,

$$c = \frac{1}{\sqrt{\mu_0 \varepsilon_0}} = 3 \times 10^8 \text{ m/s}.$$

This speed is considered instantaneous, which can extend into the question of causality and instantaneity. Being that the speed of light is a constant speed that no other known object could travel at, it was imposed that no physical speed shall ever surpass the speed of light.

But according to the physics of Newton, if a walking man is carrying a lantern emitting light, the speed of the emitted lantern light would have to be lightspeed plus the walking speed. Despite the convention of Newton's laws, it was proven through lasers and the rotation of the Earth that the speed of light does not change at all. The speed of light is still a constant, but how?

According to Einstein, both space and time (thought to be absolute in Newton's theories of motion and gravity) must change itself to satisfy the constant nature of lightspeed. Under "special relativity," space must contract, and time must dilate to conserve the physics of motion close to the speed of light. Because none of us travel at or close to the speed of light, how could we know this to be true? Let's play with lasers.

The very experiment that tested the speed of light under "Galilean relativity" was the Michaelson-Morley experiment, which hoped to prove that the speed of light changes in two scenarios: with a laser pointing in the direction of the Earth's rotation (expecting light to speed up) and with the laser pointing against the Earth's rotation (expecting light to slow down).

As previously stated, both scenarios showed that the speed of light stayed the same. The experiment provided a "null result," a nevertheless groundbreaking discovery without proving the original hypothesis. Therefore, the speed of light became a constant for Einstein's relativity.

Albert Einstein

By 1907 in Prague, Einstein would begin to see the fault in his revolutionary theory of special relativity. Special relativity only considered constant speeds, but the rotation of the Earth and the rate of free-fall depends on acceleration, a change in velocity (speed is the magnitude of velocity, which provides both magnitude and direction). It bothered Einstein until he came across the same realization Newton made; free-falling and gravity are the same thing. Instead of falling apples, Einstein thought of falling elevators.

If the suspension cables were cut and the occupied elevator would be in free-fall, the people, their carry-ons, even the elevator would fall at the same rate. They would be suspended weightless in space much like a free-floating asteroid. If you would even observe a shining laser traveling in the elevator, both while you're inside it and outside it, you'd see two different observations of the same occurrence. From the inside perspective, the laser would move in a straight line and hit the opposite wall. From the outside, however, you would see that the straight-line path of the laser is actually curved.

But how is bent light related to space and time? Space and time must change inversely to satisfy the constant speed of light. If light is bendable, then space and time are also bendable. This was the keystone to Einstein's theory of *general* relativity: the generalization to special relativity that delved into the redefinition of gravity.

Albert Einstein may be theorizing general relativity for his own purposes in generalizing special relativity. But Einstein supposed that it may solve a critical error in Newton's theory of gravity: Mercury's eccentric orbit. Mercury, the planet closest to the Sun, has a unique orbit that precedes chaotically as it makes a full revolution. Such a situation in Uranus' orbit led to the discovery of Neptune, so Newtonian physicists claimed there must be another planet closer to the Sun called "Vulcan."

After failed attempts to detect Vulcan, many believed that the Sun's brightness was concealing the planet. But Einstein was known to be a renegade and challenged the planet's existence. In 1915, ten years after special relativity, Albert Einstein published general relativity, which came with the accompanying equation:

$$G_{\mu\nu} = \frac{8\pi G}{c^4} T_{\mu\nu}.$$

Here, $G_{\mu\nu}$ describes the geometrical bending of space and time, which is related to the relativistic gravitational flux of a celestial object given off by $T_{\mu\nu}$. As we can see, both G and c are paired together in this equation. Therefore, general relativity is indeed an equation for relativistic gravitation.

The equation to general relativity is actually a network of 10 equations: ten ways to warp three dimensions of space and an additional dimension of time. Therefore, Einstein's equations are called the "field equations."

The mathematics that led to the derivation are quite intricate and complex, but the physical description to this equation answered a question about gravity Newton couldn't answer before. On the question of what gravity even is, Newton left it ambiguous and said, "the deity endures the absoluteness of space and duration," practically saying "I don't know, but God must, so there."

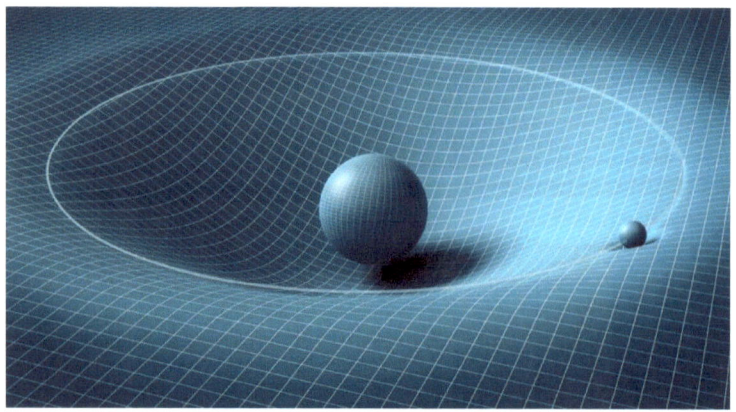

The idealization of Einstein's general relativity.

But through Einstein, gravity is induced by the bending of space and time, making gravity part of space and time, itself. Einstein even dared to say that space and time can be unified into a "spacetime continuum:" an infinite grid-like, flexible fabric warped by celestial objects to simulate gravitational fields. Therefore, gravity is not a force as Newton suggested, but rather a fundamental property of spacetime and of astronomical masses. The deeper the spacetime curvature induced by a source object, the stronger its source gravity. Orbital motion, according to Einstein, is the continuous drifting along these spacetime curvatures.

What Einstein also described through the field equations is that space and time are warped to the degrees of time dilation and length contraction from special relativity, based on the strength of an object's gravitational field. The strength of gravity is therefore related to the degree of curvature a mass applies on the spacetime continuum. Under stronger gravitational fields (steeper spacetime curvatures), a free-falling object experiences slower time and shorter lengths.

On the question related to gravitational flux, how much mass is required to even have gravitational fields,

Einstein would say *any* mass, large or small, could curve spacetime and thus have gravitational fields. Any mass can be infinitely large, but how infinitely *small* must a mass be to bend spacetime? That would be answered by Stephen Hawking with his work on "black hole entropy." But before I even mention Hawking, let me introduce you to a man who first thought up of black holes: a German soldier on the Russian front in World War I.

BLACK HOLES AND GRAVITY WAVES

It was just a mere month after Albert Einstein derived his field equations for general relativity. The man who redefined physics believed his work was done. Little did he know it was surely not the case. Albert thought his equations were unsolvable. How could anyone find the time to derive an exact solution for ten equations? General relativity was only derived to 1) generalize his 1905 theory of special relativity, 2) provide a correction to Newton's force of gravity, and 3) solve the mystery of Mercury's eccentric orbit. Einstein never expected anyone to solve these equations for an arbitrary object, especially at such short notice.

Karl Schwarzschild

The world-renowned physicist was in Berlin when he received a letter. It was from a Karl Schwarzschild: a fellow physicist, astronomer, and a soldier on the Russian front. Schwarzschild was able to find a solution to Einstein's field equations - for a supermassive, non-rotating object whose gravitational attraction is inescapable, even to light. Einstein was in disbelief: someone solved his equations, and the result is a monstrous object "not even God could have created it." Karl Schwarzschild was the one man who solved Einstein's equations for a black hole.

Because spacetime is warped by astronomical objects, these degrees of warping indicate the slowing of time and compression of space. What if time were so slow, it was "permanently frozen?" What if space were so compressed, it would become smaller than a pinpoint?

General relativity also suggests gravity redshifts light (gravity elongates the length of a light beam). Because of the extreme warping of spacetime from these black holes, visible light is extremely redshifted until it becomes invisible radio waves. The questions become the following, what is the radius of this ungodly object? How far from a black hole's center does time freeze and space compress into a singularity? And most importantly, what are black holes?

Black holes were once stars with masses of at least three suns (the mass of the Sun is 2×10^{30} kg). Firstly, a star is generated by the balance between its gravitational force and the nuclear force it produces in its core and inner layers. This is called the "Chandrasekhar limit." As stars grow and increase in mass, the nuclear energy within must maintain this balancing act with the gravitational force. That is, until the nuclear stockpile inside runs out, when the star begins its demise.

Stars have different masses, so their probability in becoming a black hole is dependent thereof. However, for massive stars of over three suns, the star's gravitational force finally overcomes its nuclear force. All atoms and particles in the star's innards are overpowered, ripped to shreds by the overwhelming gravitational force. Everything that once made up the star is now fixated within a black hole singularity. The gravitational field at the singularity is so strong, it stretches visible light into invisibility. This region of invisible "blackness" spans across the black hole's diameter of $d = 2r_S$, where

$$r_S = \frac{2GM}{c^2}$$

is the "Schwarzschild radius:" the radius of a black hole of mass M.

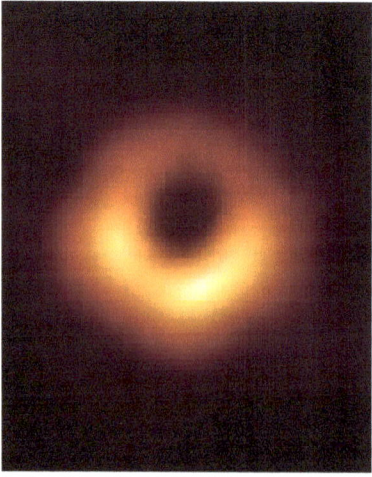

A black hole (Sagittarius A)

Solving for the escape velocity of a black hole would turn out to be the speed of light, itself. Escape velocities are generally the minimum of how fast an escaping object must move. But if the minimum speed to escape a black hole is the speed of light (and the laws of physics forbid surpassing the speed of light), then black holes are truly inescapable.

The outer edge of the black hole is called the "event horizon:" the moment in spacetime where free-falling objects appear to be suspended in space and time. Because the black hole's spacetime curvature is extreme, a black hole freezes time and compresses space into a singularity. An object free-falling into a black hole is seen by external observers to be frozen in space and time along the event horizon, while in the free-falling perspective their physical being is redshifted into a thin spaghetti noodle.

However, by 1916, one year after general relativity and the hypothesis of black holes, Albert Einstein thought of another consequence to his theory. Einstein was able to describe gravity as the inward curving of the spacetime continuum. However, if spacetime can curve inwards, it can curve outwards and send ripples like a water droplet in a calm pond. This is the basis of gravitational waves:

seismic ripples upon the spacetime geometry, gravitational surges in the 4D spacetime.

According to Einstein's field equations, an astronomical object applies a "pseudo-pressure" upon the spacetime fabric. Let's say, for example, the Sun immediately fused out and disappears. The Sun curves spacetime as it emits light. For the curved spacetime to return unbent, the continuum would ripple like a pond sending waves after a small droplet hits the surface. By the time you on Earth notice the Sun fuse out, you would also notice the Earth is no longer in orbit and is now in free float in the vacuum.

However, since astronomical masses do not disappear randomly, spacetime would have to ripple through other means: unstable black hole binaries being one of them. An unstable black hole binary is a system of two black holes spiraling inward, towards their shared center of mass. Because of their extreme curvatures, the spiraling of the two black holes sends out ripples like running your arms across the water. Once the two black holes merge, they send out the most intense pulse in the rippling, emitting out a seismic wave of spacetime. This geometric seismic wave is the gravitational wave.

Like all other waves (such as sound waves and electromagnetic waves), gravitational waves are mathematically described as a wave equation. Being that these waves are relativistic, they generally take the form of

$$\left(-\nabla^2 + \frac{1}{c^2}\partial_t^2\right)\psi = 0,$$

where the function ψ is the gravitational wave. The speed of the gravitational wave is also the speed of light, which makes gravity waves somewhat similar to light waves.

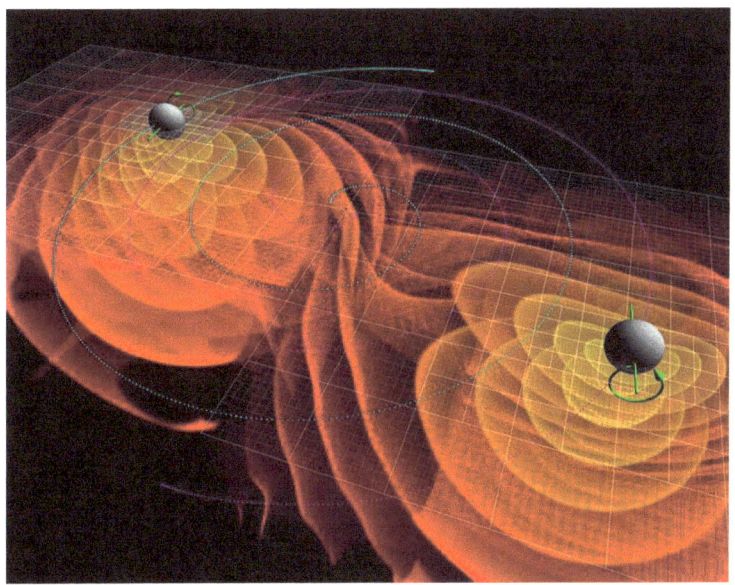

Gravitational wave formation by a black hole binary.

Gravitational waves, until 2015, were only a theoretical prediction born through Einstein's math. But once the black hole was acknowledged in 1964 by British mathematician Roger Penrose, gravitational waves seemed promising. By the 1970's, the LIGO (Laser Interferometer for Gravitational wave Observation) team was created, headed by MIT experimentalists Barry Barish and Rainor Weiss and Caltech theorist Kip Thorne. LIGO would have to detect gravity waves through super sensitive laser interferometers, a recreation of the Michaelson-Morley experiment in the large scale.

The LIGO Detectors at Livingston, Louisiana, USA.

The two LIGO detectors in Louisiana and Washington, USA, are L-shaped laser refractors, each leg being an evacuated chamber pipe spanning 4 km (2.5 miles) long. With a laser as strong as 1MW (enough power for a thousand households), these detectors must pick up gravity wave signals that shift the laser by 10^{-18} meters (19 decimal places off of zero). In other words, gravitational wave detection is an elaborate and painstaking process that demands immense precision. On September 14, 2015, the delicate LIGO detectors accomplished this feat. This won Thorne, Weiss, and Barish jointly the Nobel Prize in 2017.

Theoretically speaking, any two masses in orbit or in a binary can ripple spacetime into waves – even the Earth and the Sun. However, due to the weakness of gravity through the gravitational constant G and the smaller masses of the Sun and the Earth, these gravity waves are weak in comparison to the gravity waves of two supermassive black holes or very dense neutron stars. Such waves resonate energies within the range of 10^{46} J for neutron stars and 10^{47} J for black holes (between 47 and 48 digits of energy), fifty times stronger than any energy emitted in the universe.

In understanding gravity waves, we would be a step closer to understanding the big bang, itself, and perhaps the wave-particle duality with the graviton.

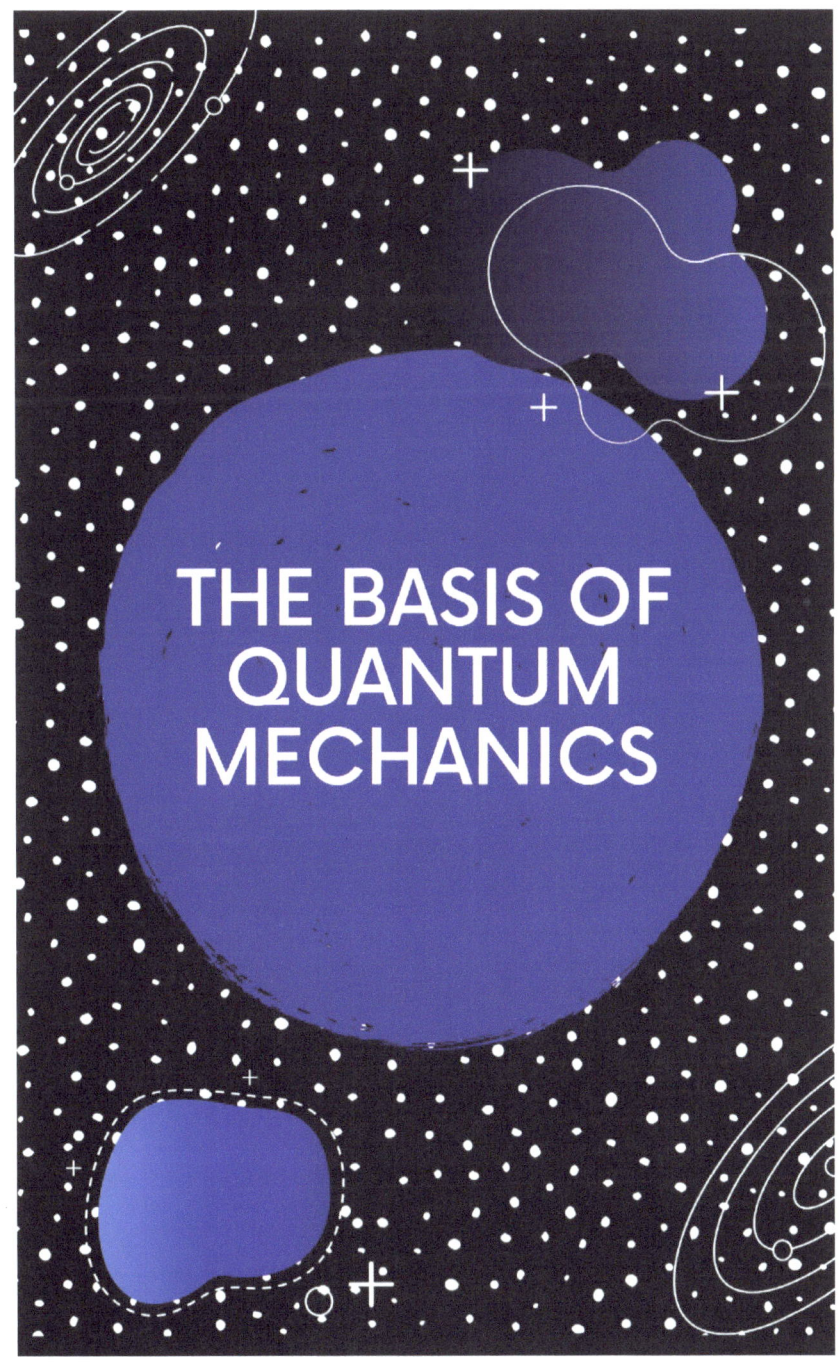

THE BASIS OF QUANTUM MECHANICS

WAVE-PARTICLE DUALITY

It was 1900 when German physicist Max Planck rectified a problem classical physics could not answer. This problem involved blackbody radiation, a blackbody being a thermal conductor that takes in heat (thermal energy) and emits light of the corresponding energy. In the classical framework the frequency of the emitted light would exponentially increase, should the blackbody heat up. This is much like looking at stove coils glowing from dark red to a light orange as heat is applied.

So thermal energy is related to the energy of the emitted light. But something happens completely different when light frequencies reach the ultraviolet region. The classical theory says the thermal energy would have a proportion to the cube of the frequency: $k_B T \propto v^3$ (higher frequencies would mean extreme thermal energies), but when the frequency is between 10^{15} and 10^{18} Hz (which is the ultraviolet range), there is no visible light in proportion to the thermal energy. This is obvious now, but back then this was revolutionary. This was called the "ultraviolet catastrophe," which was a serious issue to classical physics in the modern age.

When Max Planck solves the problem, he discovered that the relationship between heat and light is not a power rule, but a distribution. As gas particles in a box have a distribution based on their thermodynamics, emitted light shares the same distribution! However, thermal energy is mostly involved with the excitations of gas particles. Giving light a thermodynamic distribution would imply that the energy of light would contain a particle of light to excite. This would lead Planck to hypothesize that light (as well as any other form of radiation) contains discrete energy packets, or "quanta."

It was only a theory until 1905 when Albert Einstein wrote his first of the four "annus mirabilis" papers on the photoelectric effect (based on an experiment conducted in 1900): for light to ionize metals (to be able to kick electrons out of a metal sheet), there must exist a particle of light to interact with the electrons in the sheet. As the electron is emitted with energy $E = eV$, there exists "cathode radiation" that shares the same energy as the energy of the initial light:

$$eV = h\nu,$$

where $h = 6.626 \times 10^{-34}$ J*s is called Planck's constant (a constant to quantum mechanics), which assigns quantum particles a wave-like nature, and vice versa.

Because of the photoelectric effect, Planck's theory of energy quanta was made a scientific reality. This would be the birth of quantum mechanics, as well as a particular phenomenon: light (first acknowledged as a wave) would have to be a particle to interact with electrons, while electrons (understood to be particles) behave like radiation after being kicked out from the metal sheet.

This phenomenon is called the "wave-particle duality," which not only solidified the nature of cathode rays, but also led to the discovery of the light quanta called "photons:" particles of light waves. In the 1920's, two physicists would contribute further to the wave-particle duality of radiation/matter: a French Ph.D. student named Louis de Broglie and an American professor named Arthur Compton.

Louis Victor Pierre Raymond de Broglie did his Ph.D. thesis on the "quantum theory of electrons" in 1924, which led to suppose that subatomic matter in motion (such as electrons in a current) would act as a kinetic wave. This main result of the thesis would be an equation for the wavelength in relation to a particle's momentum:

$$\lambda = \frac{h}{p}.$$

This results to an equation for quantum momentum:

$$p = \frac{h}{\lambda} = \hbar k,$$

where $\hbar = h/2\pi \approx 10^{-34}$ J*s (which can be referred to as h-bar) is still Planck's constant, but reduced, and k is called the "wave number," related to inverse change of displacement of the wave.

A year earlier in 1923, Arthur Compton noticed how x-rays lose energy when radiating through crystals. Treating this as a collision problem, x-rays through crystals must act like particles colliding with the stationary electrons to transfer energy out of the refracted x-ray. He, too, would solve for a type of wavelength. But instead of kinetic particles, Compton looked at stationary "rest" particles:

$$\lambda_C = \frac{h}{mc},$$

where m is the mass of a resting particle. As de Broglie connected the momentum of a moving particle with the wavelength of a kinetic wave, Compton connected a particle's "rest energy" ($E = mc^2$ from Einstein) with the wavelength of a standing wave, calling it the "Compton wavelength." This would allow us to assign any wave a mass, as if it were a particle. Just as you can connect the de Broglie wavelength with quantum momentum, you can connect the Compton wavelength with quantum energy:

$$E = h\nu = \hbar\omega,$$

where ω is called the "angular frequency," related to the inverse change of time of the wave.

Knowing the momentum and the energy of a wave-particle is essential to another core concept of quantum physics: the conservation of energy.

SCHRÖDINGER EQUATION

Ninety-three years before Schrödinger, Irish mathematician William Rowan Hamilton postulated a form of classical mechanics that not only satisfied Newton's laws, but also looked at energy as momentum dependent. In 1833, Hamilton supposed the "Hamiltonian formalism," which supposes that the total energy of a single particle ultimately depends on its momentum and its potential energy:

$$E = \frac{p^2}{2m} + V.$$

The first term is the kinetic energy and the second term is the potential energy. Note that the right-hand side of this equation, itself, is called the *Hamiltonian,* labeled as *H*.

The competitor to Hamilton's framework of particle energy was that of the Italian-born French mathematician Joseph-Louis Lagrange, who though that the energy of a sporadic, excited particle must be mathematically reduced to find its exact physics. Instead of adding kinetic energy with potential, Lagrange *subtracted* the two energies, restraining the particle to have least action. However, Lagrange's formalism better connects Newton's physics than does Hamilton's formalism.

But as the framework of particle mechanics shifted from classical to quantum, applying the wave-particle duality would mean we would look at particles as waves. In 1926, Austrian physicist Erwin Schrödinger considered the developments of wave-particle duality and considered the *Hamiltonian* and total energy as a way to describe the quantum conservation of energy.

Erwin Schrödinger

In classical physics, measurements are based on exactness; we would know both an object's location and its speed. In quantum physics, with particles acting like waves, you cannot really tell where the particle exactly is if the wave is infinitely distributed and moving really fast. So, in the quantum world, physics must be measured with probability. This is called the "Heisenberg uncertainty principle:"

$$\Delta x \Delta p = \Delta E \Delta t = \frac{\hbar}{2},$$

which was theorized and formulated by German physicist Werner Heisenberg in 1927. If you know how fast the particle is moving (or how much energy it has) with 99% certainty, you are 1% certain in knowing where the particle is located (or its change of time).

Consider looking at very fast uniform waves with an acknowledged constant wavelength and frequency (such as light). As we let the wave pass by, all we can possibly measure is its speed and kinetic energy (just like how the LIGO detectors measured gravity waves). However, if we want to know the wave's wavelength and frequency instead, it is required that we freeze a time frame (so time is now known) and measure the wavelength. But with the frame frozen, it would be hard to tell whether the wave is really stationary or in motion. With a large uncertainty in its speed with a known wavelength, how is it possible to find the frequency of the wave?

*Guitar strings; certain strings are in wave forms that look
hazy, as if they have uncertainty.*

This is the basis of Schrödinger's contribution to quantum
physics, and so he looked at the Hamiltonian formalism.
This is perhaps why the Hamiltonian is mostly favored
across quantum mechanics rather than Lagrange's
formalism (called the Lagrangian). It's bad enough that we
don't know the actual physics of particles we cannot see.
Why reduce its energy if you're trying to decipher their
mechanics based on probability?

Looking at the Hamiltonian formalism for total energy, both momentum and energy must imply Heisenberg's uncertainty principle. Given a wave with a function ψ (as we saw with gravitational waves), momentum and energy would have to become "operators," mathematical tools dictating an applied function how to operate: $p = -i\hbar\nabla$ for the uncertainty in position (with i being the "imaginary number," or the square-root of minus one) and $E = i\hbar\partial_t$ for the uncertainty in the time frame.

Replacing momentum and energy with their operator form into the Hamiltonian, this would result to the "Schrödinger equation:"

$$i\hbar\partial_t\psi = -\frac{\hbar^2}{2m}\nabla^2\psi + V\psi,$$

which looks similar to a wave equation we saw back when we mentioned gravity waves. The right-hand side would also be called the *Hamiltonian operator*, just like in the classical theorem, denoted by $\hat{H}\psi$.

The simplest case of the Schrödinger equation is a "free particle:" a particle unbound to a potential energy ($V = 0$), whose wave function is time-independent (we focus on how a wave is distributed across all space). This changes the equation to

$$E\psi = -\frac{\hbar^2}{2m}\nabla^2\psi,$$

with the energy E being specific to the wave and to the Hamiltonian of the system.

The one issue with Schrödinger's equation is that it does not consider Einstein's relativity (it is non-relativistic). Should gravitational waves be applied into this equation, implying there is a quantum particle of gravity, then the Schrödinger equation must be relativistic:

$$mc^2\psi = -\frac{\hbar^2}{m}\left(\nabla^2 - \frac{1}{c^2}\partial_t^2\right)\psi.$$

This is called the Klein-Gordon equation (drafted by Oskar Klein and Walter Gordon in 1926). This equation certainly works for electromagnetic waves and photons: quantum particles of light. However, this equation cannot work for gravitational waves and these *gravitons*.

There is one crucial dictator across quantum physics that is unique for every particle imaginable, which even restrains a particular particle to be used in certain equations (such as gravitons and the Klein-Gordon equation). This is called *quantum spin*. An analogy to look at spin is a rotating planet in an orbit.

TOTAL ANGULAR MOMENTUM

Consider the Earth orbiting around the Sun. While the Earth is in orbit, it is also rotating about its axis. The Earth's orbit has an "orbital" angular momentum L, depending on how fast its orbital speed is and how far the Earth is from the Sun. The planet's rotation has an "intrinsic" angular momentum S, which is only dependent on its magnetic field induced in its core (which is different for other planets). So, the Earth would have a total angular momentum $J = L + S$, which would only increase for astronomical masses.

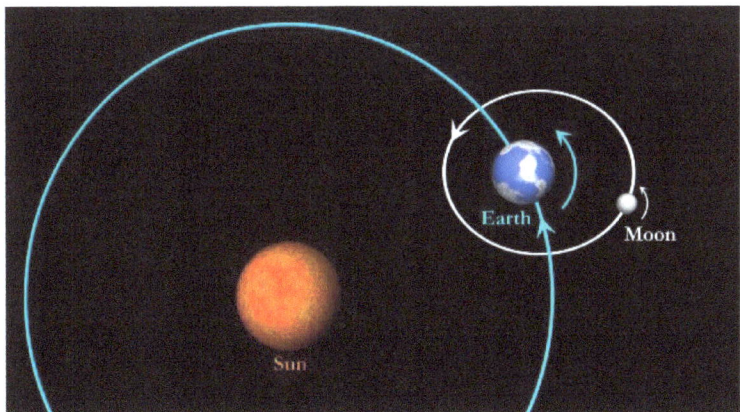

The Earth's rotation and orbit.

For quantum particles, the situation is the same, but added with a small revision: the total angular momentum would either increase (if L and S are pointing in the same direction) or decrease (if L and S are pointing in the opposite direction). But instead of an orbit, let us look at a well. Just like a water well plastered by endless layers of brick and mortar, a "potential well" is layered by numerous energy states: energy excitations of a particle inside this well. The higher the state, the energetic the particle. As the energy state increases, the more likely a particle would gyrate with translational energy (moving side to side).

If we look at electrons in the atom, their energy states depend on their orbital motion around the nucleus. If any particle is inside a potential well, their energy states are related to various orbitals around the node of the well. From a side view, the particle in any higher energy state would gyrate translationally. From a bird's eye point of view, this translational motion is orbital motion. This means that particles in any energy state have an orbital angular momentum:

$$L_z\psi = m_l\hbar\psi \qquad\qquad |L|\psi = \hbar\sqrt{l(l+1)}\psi,$$

where l is the azimuthal number (always a whole integer one value less than the energy state; at the ground state, the azimuthal number is zero) and m_l is the "magnetic azimuthal number," a value between $-l$ and $+l$. Already, we can tell that quantum angular momentum can alternate from positive to negative, and this alternation is (like the rest of quantum mechanics) probabilistic.

As a planet's rotation is independent of its orbit, so is a quantum particle's intrinsic properties. This property is called *spin:*

$$S_z\psi = m_s\hbar\psi \qquad\qquad |S|\psi = \hbar\sqrt{s(s+1)}\psi,$$

where s is the spin number and m_s is the "magnetic spin number," a value between $-s$ (spin down) and $+s$ (spin up). Unlike the azimuthal number l, the spin number can either be a whole or half integer, which also determines the class of quantum particle.

Quantum particles with half integer spin numbers are called *fermions* (named after Italian physicist Enrico Fermi), which follow a unique distribution called "Fermi-Dirac distribution," and a set of restrictive rules such as the Pauli exclusion principle (you cannot have more than two fermions in the same energy state). Examples of fermions are the electrons, protons, and neutrons, which all have the spin number of $s = 1/2$.

Quantum particles with whole integer spin numbers are called *bosons* (named after Indian physicist Satyendra Bose), whose quantum mechanics are more relaxed. Under their own unique distribution called "Bose-Einstein distribution," bosons are free to occupy any energy state, as numerous as they can. Examples of bosons are the photons (which has a spin number of $s = 1$) and the gravitons (which has a spin number of $s = 2$). This is why the Klein-Gordon equation cannot work for gravitons. The Klein-Gordon equation only satisfies spin-1 particles, specifically photons. Because gravitons are spin-2 particles, their quantum mechanics requires an intricate "spin-dependent" approach.

For all quantum field theories, there exist a "field boson" and a corresponding "fundamental fermion." For electromagnetism, the photon (light particle) is the field boson, with the electron being the fundamental fermion. For the strong nuclear interaction, protons and neutrons are held together by these fermions: the up and down quarks; the field boson is the gluon (the glue between the quarks).

And for the weak nuclear interaction, nuclei fizz out and split between smaller nuclei and "neutrinos;" the weak interaction has two field bosons: the W (weak) boson for mediating neutrinos; and the Z (zero-charge) boson to conserve the momentum, spin and energy of the fission process. In the absence of gravity, this zoo of bosons and fermions form the Standard Model, a theory of uniting electromagnetism and the nuclear forces into one coexisting interaction.

For quantum particles with orbital angular momentum *and* spin, there exists a total angular momentum:

$$J_z\psi = m_j\hbar\psi \qquad\qquad |J|\psi = \hbar\sqrt{j(j+1)}\psi,$$

where $j = l + s$ is the total (coupled) angular momentum number and m_j is the "magnetic coupled number," a value between $-j$ and $+j$.

As we approach discussing the graviton and gravity as a quantum theory, it is important to acknowledge spin (especially for gravitons with spin-2 properties) as well as the coupled angular momentum (especially when discussing string theory and loop quantum gravity). The previous quantum chapters on wave-particle duality and the Schrödinger equation will be tied with the past chapters on classical gravity, in order to unlock the secret of a quantum field theory for gravity.

HAWKING RADIATION

Hawking radiation applies quantum mechanics to astrophysics, especially when we look at black holes. The theory was drafted by Stephen Hawking in 1974; fifty-nine years after general relativity and the hypothesis of black holes, and forty-seven after the birth of quantum mechanics. As the age of classical physics came to an end, quantum physicists were desperate in redefining the classical field theories of electromagnetism and gravity into *quantum field theories*.

In classical field theory, an interaction of electricity or of gravitation is induced by a field, either the electric field or the gravitational field. For electrodynamics, to be specific, the electric field is changed in such a way that light is emitted. In quantum field theory, an interaction of electrodynamics is induced by an *electron* colliding with a *positron* to produce a *photon*. In this case, the electron and the positron (a particle-antiparticle pairing) collide and "annihilates" into a particle of light. This conserves the energy, charge, and quantum spin of the interaction.

These particle-antiparticle pairings are crucial in the quantum sense of space: space is not really empty and composed of nothing as we may think of it to be. Space is actually jittering with quantum excitations that "stabilizes" space by these annihilations. This is called the "Casimir effect," which actually skims the surface on one of the quantum field theories of gravity. Because particle-antiparticle pairings stabilize space, this is completely turned upside-down when considering a black hole.

Recall that black holes are sources of extreme spacetime curvatures where time is frozen, and space is squeezed into a singularity. Inside a black hole, space is no longer "space" anymore. This means, in the quantum sense, particle-antiparticle pairings cannot stabilize whatever is inside a black hole. But, according to Hawking, the only way a black hole may stabilize itself is through a form of quantum radiation. In modern quantum physics, particles contribute to everyday matter (electrons, photons, even the soon-to-be-mentioned gravitons); antiparticles contribute to anti-matter, which has an incredibly short lifespan in exposure to "normal" matter.

In regard to black hole stabilization, a particle-antiparticle pairing is "ripped apart" at the event horizon, where the normal matter particles are emitted outward away from a black hole, leaving the antiparticle inside it. This emission of normal particles and cumulating of antiparticles inside a black hole both contribute to the black hole losing its mass. Quantum bits of radiation is emitted, shaving away a fraction of the black hole's total mass.

Stephen Hawking

Theoretically speaking, as more and more particle-antiparticle pairings stored in a black hole are split apart, that black hole may evaporate into a "quantum black hole," where all the antiparticles are gyrating with immense energy inside such a closed volume. This scenario is quite similar to thermodynamics, where the energy of the particles inside a space depends on the pressure, volume, and temperature of the enclosing. In other words, this collection of particle-antiparticle splittings also contributes to black hole thermodynamics.

This radiation of normal particles from a split pairing is called *Hawking radiation,* which can also be thought of as thermal radiation of a black hole. Hawking considered, the entropy of a black hole is related to the ratio between its surface area and the square of the smallest radius it may ever achieve:

$$S = k_B \frac{4\pi r_s^2}{(2l_P)^2},$$

where $k_B = 1.38 \times 10^{-23}$ J/K is the Boltzmann constant (a constant of thermodynamics), and $l_P \approx 10^{-35}$ m is the "Planck length," the smallest length to possibly achieve.

To help visualize the smallness of the Planck length, let's say the entire Milky Way galaxy can fit in the Earth from edge to edge. A Planck length is roughly the size of your living room.

Therefore, the radius $2l_P$ is the Schwarzschild radius of a quantum black hole, making its mass the "Planck mass" of value $m_P = 2.18 \times 10^{-8}$ kg (roughly the mass of a flea egg). According to Hawking, this mass must be the smallest mass to curve spacetime, which would make sense since electrons and photons are negligible in mass to even have their own gravitational fields. Squaring the Planck length would lead to a unique combination of constants:

$$l_P^2 = \frac{G\hbar}{c^3}.$$

With G being related to gravity, c related to Einstein's relativity and $\hbar$ related to quantum mechanics, Hawking radiation and black hole entropy is a relativistic quantum gravitational interaction, just as it is essentially thermodynamic.

Through black hole entropy, one can determine the actual temperature of a black hole, which is related to the storing of antiparticles and its energetic gyrating inside:

$$T = \frac{\hbar c^3}{8\pi G M k_B}.$$

The larger the black hole, the colder it is. The smaller, the hotter. When black hole temperature is applied to Hawking's Ph.D. thesis, we could even determine the temperature of the big bang. Letting the mass of the universe at the beginning be the Planck mass, the temperature would turn out to be the "Planck temperature:" $T_P \approx 10^{32}$ K.

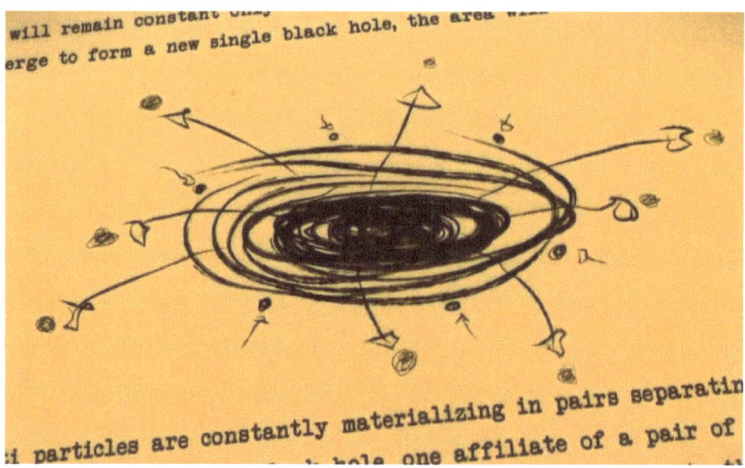

Hawking radiation depicted in the original paper.

In the general sense, Hawking radiation gives us an insight what is inside a black hole, aside a singularity and a point of no physical return. Black holes are not really black at all, but "glow" with heat radiation; and are not empty as we would expect them to be. Perhaps the evaporation of black holes may even tell us what space is comprised of in the fundamental core. This may tell us what the quanta of spacetime is, as well as the quanta of gravity.

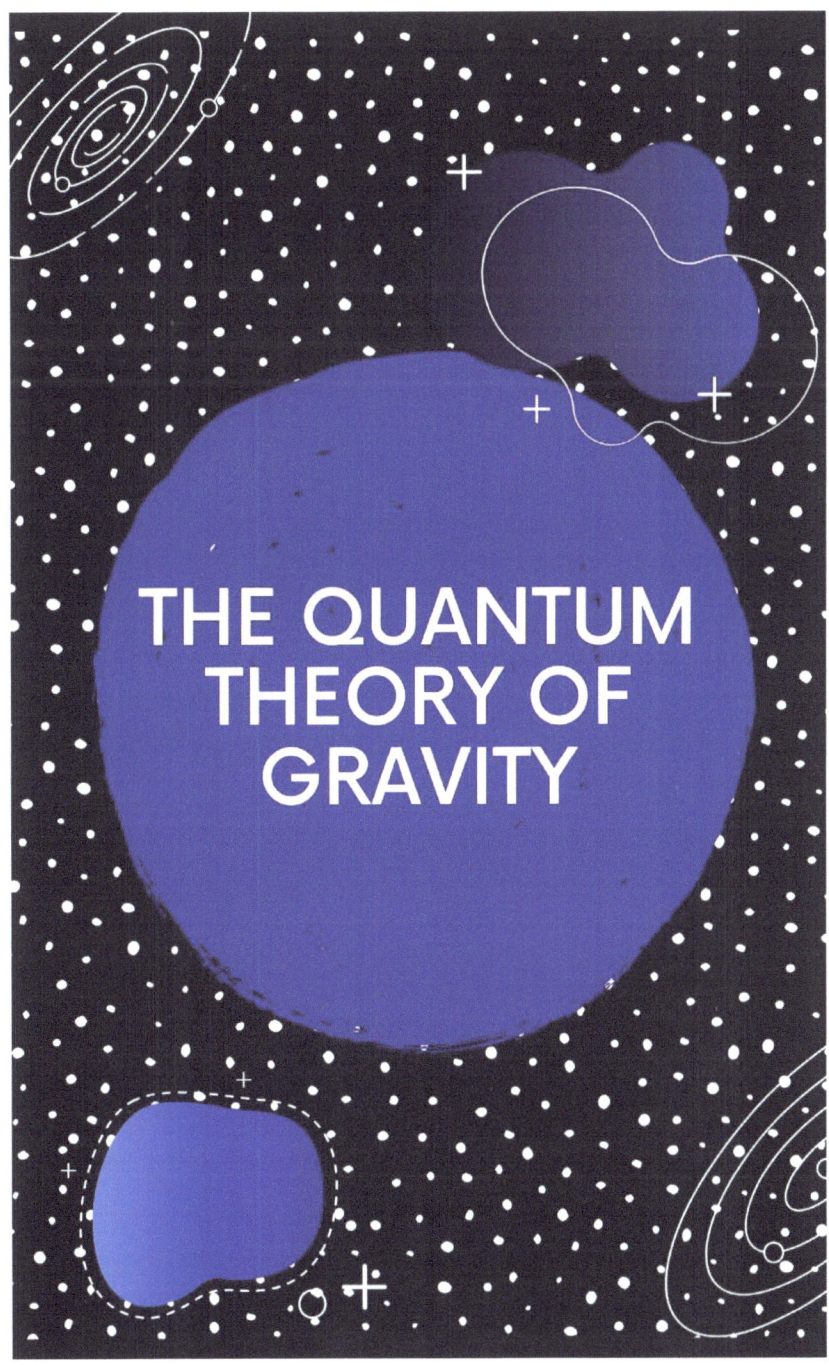

THE QUANTUM THEORY OF GRAVITY

NUMEROLOGY THEORY

Numerology theory is the theory in which the combinations of physical constants and their numerical values help define the fineness, or the order of magnitude, of a given physics problem – and from which the creation of more fundamental constants of physics. Maxwell used this theory to find the speed of light from the constants of electricity and magnetism. Also, English physicist Paul Dirac, the man behind the relativistic quantum theory for electrons, used numerology theory to numerically discover the proton's mass and the "fine structure constant" for quantum electrodynamics:

$$\alpha = \frac{e^2}{4\pi\varepsilon_0 \hbar c} \approx \frac{1}{137}.$$

This seemingly random number of 1/137 comes from a mathematically specific combination of constants, especially those of electrostatics (e is electric charge and ε_0 is electric permittivity), quantum mechanics $\hbar$ and Einstein's relativity c.

It does not hurt to suppose, numerology theory can help look into the quantum theory of gravity. I must warn, however, the math can get messy.

Suppose we have a hydrogen atom. In it, an electron and a proton have gravitational attraction just as they have electric attraction:

$$\frac{1}{4\pi\varepsilon_0}\frac{e^2}{r^2} = G\frac{m_e m_p}{r^2},$$

which considers that the ratio between electric charge and particle mass is an electro-gravitational constant connecting gravitational attraction to electric attraction:

$$\frac{e^2}{m_p m_e} = 4\pi G\varepsilon_0.$$

In the hydrogen atom, the electric attraction between the electron and the proton is much, much stronger than their gravitational attraction by 44 digits. However, theoretically, a displacement infinitely far away can make the gravity and electricity forces equal. The question is: how far away is that the case?

From classical theory, the charge to mass ratio of an electron was tested on and calculated by J.J. Thomson in 1897, which is based on the guiding electric field and a drifting magnetic field. Assuming the electron is on a linear path, and that the particle barely drifts at all, the electron's charge-to-mass ratio is simplified:

$$\frac{e}{m_e} = \frac{8\pi}{l^2\mu_0 e}.$$

Here, μ_0 is the permeability constant from magnetism and l is the length of the electron linear path. In order for the electric attraction to be equal in strength as the gravitational strength, we need to know how long the length l is. The electro-gravitational constant is revised further down to

$$\frac{1}{m_p}\left(\frac{1}{l^2}\right) = \frac{G}{2c^2}.$$

Using Paul Dirac's numerology of the proton mass: $m_p = 4\hbar/cr_p$, where $r_p = 0.814 \times 10^{-15}$ m is the proton radius, this solves for the electron linear path to be

$$l^2 = \frac{r_p c^3}{2G\hbar} = \frac{r_p}{2l_P^2},$$

where the Planck length l_P reappears.

Another importance of knowing the length of the electron linear path in this scenario is that the resulting number gives us the theoretical extent of the universe. The electron being drawn towards the proton moves from its outermost edge to the center of the atom. The reverse of inward motion is, of course, outward motion; and the universe is expanding outward from the epicenter of the big bang.

87

The length from the electro-gravitational interaction is calculated to be $l = 1.28 \times 10^{27}$ m, which is larger than the current radius of the universe ($r_U = 8.80 \times 10^{26}$ m). Since the Planck length is derived from our particular problem, the Planck length is the fundamental constant of quantum gravity. This is essential to reinterpret Einstein's general relativity and the gravitational field as a quantum field.

STRING THEORY

String theory is one candidate for a quantum field theory of gravitation, which looks at particles as vibrating strings. The theory was born in the 1960's, when nuclear physicists researched more into a type of fermions called "hadrons:" particles made of three quarks fastened together by the strong nuclear interaction. In nuclear physics, protons and neutrons are "nucleons:" hadrons made of "up" and "down" quarks that determine their charge and their quantum spin.

The strong nuclear force that fastens the quarks inside the nucleons can be resembled as a vibrating string, while being thought of as the "gluon" field boson (the glue binding the quarks in the nucleon). Therefore, since Arthur Compton made the earlier connection between standing wavelengths to particle mass, a particle may be envisioned as a vibrating string (a standing wave) whose vibrations indicate mass, or rest energy.

Uniquely, the vibrations upon a particle string can be described as the relativistic action upon that string:

$$S = -T_0 \int dA,$$

where $T_0 = mc^2/s$ is the string tension, s is the path of the string and dA is the surface area of a "world sheet:" a cylindrical surface traced out by the "worldline," or a particle string.

To make the string action more relatable with the length of the particle string l_s, a "slope parameter" α' is introduced. Slope, in this case, is not necessarily the literal steepness of the string in 4D spacetime, but rather the angular momentum intensity on the string (which is directly due to quantum spin). This changes the action as

$$S = \frac{-1}{2\pi\alpha'\hbar c} \int dA,$$

making

$$T_0 = \frac{\hbar c}{2\pi l_s^2} = \frac{1}{2\pi\alpha'\hbar c},$$

where $s = 2\pi l_s$ and $E = \hbar c/l_s$.

Solving the length of a particle string, l_s, helps identify the particle by its mass, charge, or quantum spin. The larger the magnitude of any physical property, the longer the string. Consider wrapping yarn around a baseball and a bowling ball. A baseball, smaller in size and easier to lift, would have a smaller circumference of yarn in comparison to the bowling ball, much larger and heavier.

From the above, the string length is therefore

$$l_s = \hbar c \sqrt{\alpha'},$$

which ultimately depends on the slope parameter. Meaning, we have to define what α' is. Unitwise, the slope parameter is the inverse-square of the particle energy. But to incorporate spin (in general angular momentum), the slope parameter must be proportional to angular momentum:

$$\alpha' = \frac{\lambda^2}{(\hbar c)^2} \sqrt{j(j+1)},$$

where the quantum number j is from total angular momentum and λ is the length eigenvalue (or a particular value for length) of a particle string, which is not always the particle's Compton wavelength. This changes the length of the string to be

$$l_s = \lambda \sqrt{\sqrt{j(j+1)}}.$$

To limit confusion between wavelength and the length eigenvalue, I change λ to l.

So, from string theory, every particle is interpreted as a string whose length is directly affected by its quantum spin. Larger spins lead to larger lengths. This introduces a new quantum operator: a length operator,

$$\hat{L}\psi = l\sqrt{\sqrt{j(j+1)}}\psi.$$

This may work well with particle physics, but how does string theory so far work with quantizing gravity? Since particles are depicted as vibrations along a string, and to which a world sheet, this world sheet can be unraveled into a "particle field," or a spacetime fabric whose geometrical ripples are the physical entity of a particle. Like ocean waves, these ripples are described by its energy and vibrational frequency, which string theory focuses on.

Every particle has its own field, and the "stacking" of different, rippling fields upon one another outlines an interaction between these particles upon spacetime. Provided the interactions of various particles in a given region, this stacked sliver of particle fields would look quite like a vibrating membrane encompassing the four dimensions of spacetime. Should an auxiliary, unbendable metric rest at the equilibrium of the particle fields, each hump surfacing from the auxiliary field would look like emerging world sheets (reduced down to linear strings) whose ends are attached to the static field. Such strings are called "open strings," whose open ends are fastened to the auxiliary field that "hold together" a set of 4D spacetime.

Suppose there is a second spacetime membrane, or "brane," on top of the first reference brane, certain particle fields can have enough energy to seep into the higher brane, linking the two sets of 4D spacetime via particle "leakage." Such particles that can leak from one brane to another are called "closed strings," whose open ends are latched to each other, unbound to any set of 4D spacetime. Leakage is quite analogous to transitioning in quantum states, where a particle can freely move between energy states.

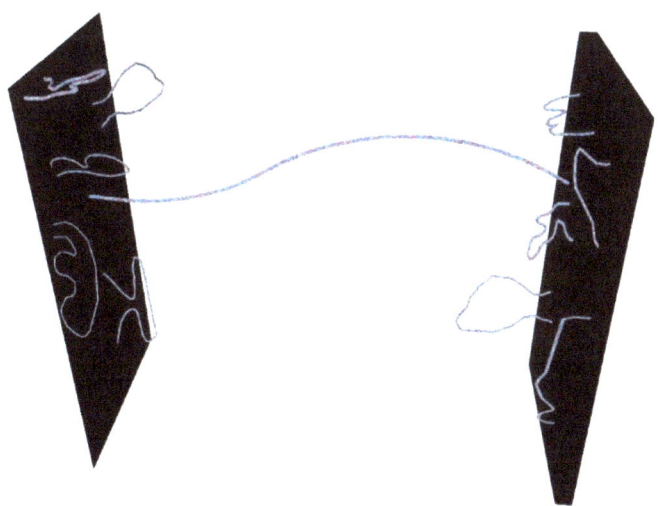

Two branes connected by a closed string (open string latched to one brane and another).

Like much of quantum mechanics, leakage is a spontaneous, stochastic event that can happen without any warning. However, we can focus on our brane, and what is happening there. This continuous "fizzing" of quantum fields emerging from the static metric is essential to the "stabilization" of spacetime in the Planck scale. This is the Casimir effect, which was mentioned when we discussed Hawking radiation. This quantum description of spacetime from string theory leads into loop quantum gravity.

LOOP QUANTUM GRAVITY

Loop quantum gravity (LQG) is the second candidate theory to gravitational quantum field theory. Instead of looking at gravity as an interaction, LQG looks at gravity as the bedrock of quantized spacetime. That is, instead of looking at brane transitioning or particle leakage, we look at how the vacuum of spacetime is resembled down to the pixel. Although the theory matured in the 1980's, the theory of quantum gravity really began in 1966, at the Raleigh International Airport.

Physicists John Archibald Wheeler and Bryce DeWitt were close collaborators in a quantum theory of gravity. DeWitt lived in North Carolina, and Wheeler always traveled to see his collaborators whenever he could. It was at the Raleigh Airport, where Wheeler waited on a connecting flight, where DeWitt presented him a mathematical equation.

Wheeler and DeWitt's main concern was expressing Einstein's general theory of relativity into a quantum equation. Albert himself wanted to do so, but sadly he died in 1955 while working on unifying quantum mechanics with general relativity. But even then, it was a mess. Quantum mechanics throws a shroud of uncertainty on a test object based on location and momentum, which both are needed in general relativity. But DeWitt approached it from another way. Instead of looking at the quantum fuzziness of a source object's location and its momentum upon spacetime, what if we looked at the quantum fuzziness of spacetime, itself? The singular fabric of spacetime thus became a *Schichttorte*, a layer cake, of sub-metrics. Each sub-metric is a slice of smaller space that each have a probability of quantum curvatures. This is called the "ADM Formalism," named after physicists Richard Arnowitt, Stanley Deser and Charles Misner.

However, to explain the quantum fabric of spacetime to be as flat as it is by default in the astronomical scale, the Hamiltonians of each spacetime slice, each resembles a Planck-scale geometric curvature, must add to zero:

$$\sum_{n=0}^{\infty} \hat{H}_n \psi_n = \frac{G\hbar}{c^3} g^{\mu\nu} \partial_\mu \partial_\nu \Psi = 0,$$

where ψ_n is an individual curvature function on each spacetime slice, Ψ is the "spectrum function:" an overlap of these individual functions into a grand function of spacetime curvature, and $G\hbar/c^3$ is the square of the Planck length.

This is the Wheeler-DeWitt equation: the mathematical equation Bryce DeWitt showed John Wheeler at the Raleigh Airport, which was initially called the Einstein-Schrödinger equation.

Until the 1980's the Wheeler-DeWitt equation was unsolvable. In the meantime, string theory was being formulized as a theory for the Standard Model. When it became possible to look at unraveled world sheets as vibrating fields, loop quantum gravity was beginning to make an appearance. In this ADM Formalism, a sub-metric that is part of the main fabric of spacetime has the same quantum fizzing as a particle field. As you stack particle fields in string theory, you have a vibrating brane that outlines quantum spacetime. To solve the Wheeler-DeWitt equation, the flat quantum spacetime must be analogous to a vibrating brane: a fabric that is "foaming" with quantum uncertainty.

Therefore, each of the Hamiltonians in the Wheeler-DeWitt equation is for a vibrating string, which in this case resembles a chunk of spacetime laced into the quantum spacetime. Therefore, each laced chunk has the length operator

$$\hat{L}\psi = 2l_P\sqrt{2\pi\sqrt{j(j+1)}}\psi,$$

where the length eigenvalue here is proportional to the Planck length – particularly the radius of a quantum black hole.

Unlike string theory, where the length eigenvalue represents the wavelength of a vibrating particle string, the length eigenvalue here represents a unit length that separates a quantum chunk from another. If looking at Albert Einstein's grid-like spacetime continuum from general relativity, a Planck length is a length from one grid point to another. Meaning, the Planck length is a side length of a square in the quantum lattice. Each of these grid squares are actually chunks of space, interwoven and fastened into a "spin network." The perimeter around each grid square is the loop behind LQG.

Like world sheets in string theory, these spin networks have an area. Here, the area of a spin network is done by coupling two length operators into an area operator:

$$\hat{A}\psi = (\hat{L} \cdot \hat{L})\psi = 8\pi l_P^2 \sqrt{j(j+1)}\psi,$$

where 8π is a number that is coincidentally from Einstein's field equations for general relativity.

As it so happens, its half integer 4π is a leading coefficient for the area of a sphere. This gives these square spacetime chunks a spherical surface area in that assumption. Since each chunk is space, they cannot be plucked or unwoven from the grid lattice.

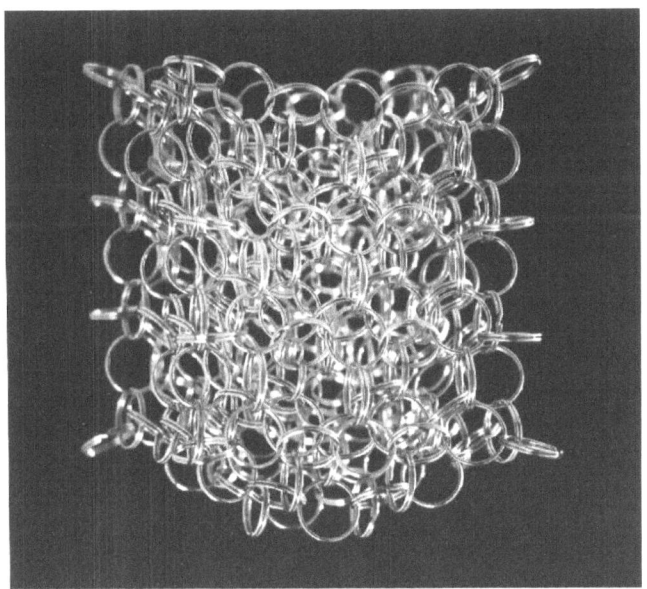

A 3D model of a spin foam in LQG. Courtesy of Carlo Rovelli.

In the context of loop quantum gravity, Einstein's fabric of spacetime is a fizzing, foaming surface. Each bump and hill come and go, at different and random positions upon the spacetime. Since each "foam bubble" appears and disappears, this is similar to the Casimir effect, how particle-antiparticle pairing appear and collide to preserve the quantum stability of spacetime.

To that extension, loop quantum gravity preserves the validity of Hawking radiation. However, direct experimentation is far from tangible. To test that quantum spacetime is indeed a foaming surface, gamma ray bursts were tested to see that the time of arrival is increased due to some form of friction from the spin networks. The accuracy of time arrival is however spot on to the calculation.

Now the new question is, *how can string theory and LQG be further compatible?* It is now time to consider the graviton: the quantum particle of gravity.

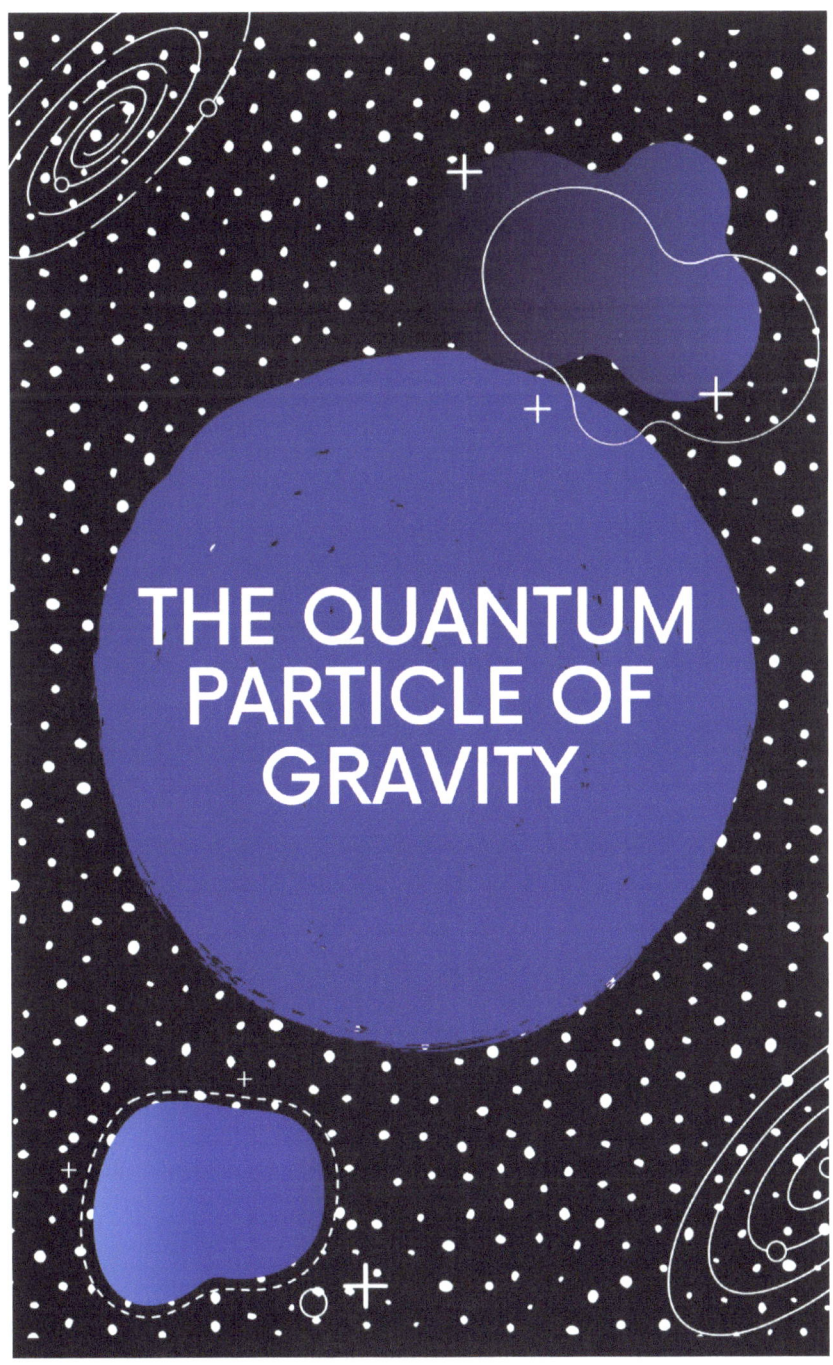

THE QUANTUM PARTICLE OF GRAVITY

WHAT ARE GRAVITONS?

The year was 1934; two Soviet physicists named Dmitrii Blokhintsev and F.M. Gal'perin were writing a report on neutrinos, which were only a theory by 1930. As a neutron decays down to a proton and an electron, an additional particle must be emitted to conserve the momentum, energy, and quantum spin of the process. This particle would be chargeless, near massless and indistinguishable, but nevertheless essential to the physical conservation laws of beta decay. These were the neutrinos, the fermions of the weak interaction.

However, the Soviet scientists would expand on this thought by applying to the other fundamental interactions (particularly gravity). From a gravitational interaction between two celestial bodies, there must exist a fundamental neutrino-like (meaning miniscule, massless, chargeless and indistinguishable) *quantum particle of gravity* that would harness the gravitational force. It was then when Blokhintsev and Gal'perin theorized the graviton.

The theory behind a particle (or an energetic entity) of gravity was not a theory born from quantum field theory. In the 1950s, the main directive of modern physics and general relativity took a rather heavily forested path with no machete in hand to clear the way. Einstein's gravity theory states that any object with rest energy can warp the neighboring spacetime. This, thus, supposes that even an object purely made of radiation can curve spacetime.

Of course, the context here is a classical object constructed by localized radiation. But the scale is rather left ambiguous. These classical orbs of energy were called "gravitational geons," which can be thought as a classical particle of gravity (independent of all sorts of quantum weirdness). However, with a paradigm shift from modern physics to quantum physics, the question of the classical geon finally had its machete and made way towards the quantum graviton.

In the very early developments of string theory, when the gluons between the quarks resembled a vibrating string of energy, physicists came across extra vibrating nodes along the gluon string. These extra vibrational nodes were a secondary particle transferring back and forth along the gluon string: a massless, spin-2 boson that behaved like photons, but resemble gravitational interaction. These bosons were the gravitons!

From this train of thought, however, the mainstream thinking from the string theorists and Standard Model particle physicists considers that the gravitons originate from the quarks and astronomical masses, themselves. This is similar to Newton's consideration that classical gravity comes from the massive objects. Since quantum gravity is about quantizing Einstein's general relativity, it would be more appropriate to suppose that gravitons somehow come from spacetime, itself, which can be excited and emitted from the massive object's source of curvature.

For there to be a successful quantum field theory for gravity, the graviton must be the field boson of the gravitational field. Should the classical gravitational field be quantized into gravitons, it must be thought that celestial bodies induce this leakage of gravitons just as they induce classical fields.

Gravitons must also be in a wave-particle duality with gravitational waves. Given that a gravity wavelength is 10^{16} m (17 digits of length), letting this be the graviton's Compton wavelength, gravitons may have a mass smaller or equal to 10^{-58} kg (59 decimal points off of zero), which satisfies the assumption of being massless. Therefore, they travel at the speed of light just like photons. Due to this similarity, gravitons have the nickname "the blue photon," blue perhaps because of the black emptiness of outer space.

General relativity describes the flexibility of spacetime, especially by means of inertial curvature or gravity wave propagation. Because of which, gravitons can, too, become "flexible" in a property called polarization. Unlike the polarization of charges (where a net neutral object can have a positive and a negative half) or the polarization of light (where white light entering a crystal refracts inside and exits out as the rainbow spectrum), polarization of gravitons involves the flexibility of spacetime due to gravitational waves (meaning the parameters of space and time alternate and oscillate like a cube of gelatin). There are two types of malleable polarization: "+" (space and time stretch inversely between a vertical oval and a horizontal oval) and "×" (the oval spacetime stretching is oblique).

As general relativity discusses gravity as the warping of spacetime due to mass and energy, gravitons must show some attraction to mass and energy. They may not have a charge or mass to be mutually attracted to anything, they are nonetheless attracted to energy. This would mean that gravitons are mutually attracted to both photons and to other gravitons, and most importantly to astronomical masses and free-falling objects to simulate gravitation.

Because gravitons have attraction to energy, it can be thought that they "self-gravitate." This would mean they, themselves, emit smaller gravitons. These smaller copies are called "daughter" or "virtual" gravitons, dubbing the source graviton the "mother" particle. This property of the graviton is neglected when looking at the quantum mechanics of just one graviton in spacetime. However, if wanting to show quantum scattering between photons and gravitons (even graviton-graviton scattering), this complex property must be considered.

For the other fundamental forces, there is a field boson with a fundamental fermion. Electromagnetism has photons and electrons, the strong interaction has gluons and quarks, and the weak interaction has the W & Z bosons and neutrinos. While the graviton is the field boson for gravity, there exists a "gravitino," a spin-3/2 fermion. However, unlike the other boson-fermion pairings (where the particles can share the same space), gravitons and gravitinos are never seen together. This is based on a sub-theory in string theory called "supersymmetry," which considers the strength of gravity in higher sets of dimensions. Perhaps, gravitinos may contribute to anti-matter or dark matter due to not being in the same observation space as gravitons.

And most importantly, gravitons are thought to have a stable lifetime and they never decay. The big bang would be considered to be the universe's first gravity wave, emitting a seismic wave across spacetime into the deepest corners of whatever lies beyond our observation. Nothing casts a shadow in front of gravity, just as you can cast a shadow in front of a light source. You live in a house, under a roof, casting off the Sun's heat and light; but you can't cast away the Earth's gravity. Gravity can exist wherever light cannot. Meaning, as photons obliterate whenever a shadow is cast or when you see color, gravitons do not obliterate when they mediate through various media. Gravitons are eternal, just as space, time and gravity are all eternal.

CURRENT THEORIES OF THE GRAVITON

Because the graviton is a hypothetical particle, many physicists over the years have proposed a variety of theories for this particle. Being that is book is partially about the graviton, I will include all prospect theories of the graviton particle, current and tentative. But first, I will discuss the current theories of the graviton that has some ounce (or Planck mass) of validity and sense of the quantization of general relativity.

There are three current theories of the graviton particle: composite theory, brane theory and black hole well theory. One of which is linked with string theory; another linked with the Standard Model. Keep in mind that the graviton must be a quantum particle in sync with Einstein's general relativity, how gravitons are not just particles of gravity, but also the particles of spacetime.

The composite theory of gravitons considers the creation and composition of gravitons. Just as how electrons and positrons collide and create a photon in quantum electrodynamics, the graviton must be a product of smaller elementary particles in a quantum gravitational interaction. This theory has the potential to be ingenious as well as terrible, for it can split into two directions.

Direction 1: two gravitinos (a gravitino being the spin-3/2 fermion partner to the graviton) may collide to produce a graviton and a photon. This conserves quantum spin and electric charge easily, but the question is whether mass (related to momentum and energy) is conserved. A graviton is near-massless with a 10^{-58} kg threshold, but these gravitinos are heavier with a 10^{-24} kg threshold (heavier than electrons *and* protons)! Perhaps these gravitinos have chiral mass (chirality being the property to have a mirror image of any physical characteristic), which implies that gravitinos have "negative" or "anti" mass either at will or conditionally. This may uncover the mystery of dark matter.

Direction 2: astronomical masses and larger quantum particles physically create and emit gravitons. This may mean that gravitons are manufactured by the quantum DNA of foreign particles (either out in deep space or in the planetary cores), implying that gravitons may be made of quark-like constituents. Ultimately, gravitons are made of a quark-gluon plasma that, as a whole, manifest into something purely gravitational. This is hogwash, because gravitons are not hadrons (fermions made of quarks), let alone fermions themselves.

These two directions of composite theory are dueling interpretations on the genesis of gravitons. Direction 1 shows merit, especially when considering the genesis of universal gravity at the moment of big bang. It also shines light on what gravitinos may represent in the present time, whether they may be particles of dark matter or of supersymmetric gravity, and what gravity is identified as in the quantum sense. However, for graviton creation in the present time, direction 2 is favored by Standard Model physicists.

But here is the absurdity of composite theory. Direction 2 completely disregards the notion of gravity under Einstein's theory of general relativity: gravity is the property and characteristic of spacetime, not of mass and energy. To say gravitons are created by other fundamental particles, while still calling gravitons fundamental, has no consistency on what it means to be fundamental. Either gravitons are fundamental or not; if gravity is fundamental to space and time, per Einstein's relativity, which might I add are more eternal than mass and energy by only a time difference of $1/(45 \text{ digits})$ of a second after the moment of big bang, then gravitons are fundamental as a particle composed of spacetime, not of foreign quantum energy.

The second theory of gravitons is brane theory, which can be seen as a consequence to string theory. Just as Albert Einstein unified 3D space with time into a 2D spacetime continuum, "quantized" spacetime can be thought of a vibrating, malleable 2D membrane (or "brane"). Because gravity is induced by spacetime warping and rippling, gravitons are practically fizzing upon the brane.

As we consider higher dimensions, more layers of branes would stack upon each other like a layer cake. And as gravitons fizz on one brane, they can theoretically have enough "fission" energy to transfer itself into a higher brane. This is because gravitons are closed-string particles, which can travel between dimensions. This is to say it requires a tremendous amount of energy to transcend beyond these four dimensions of spacetime, and that gravity is *stronger* in higher dimensions.

It shows merit in suggesting that gravity, the weakest of the four interactions, can be stronger. But how could we know that for sure? We cannot. Our observation of physics is limited to this 4D spacetime brane. Our massive scale forbids us from physically seeing what is beyond our three dimensions. Quantum particles, however, do.

Like all of quantum mechanics, brane theory is more based on probability than actual fact: the likelihood for a graviton to jump dimensions is improbable, yet possibly close to 99%. We shrug our shoulders out of intellectual cluelessness.

The third theory of gravitons is black hole well theory. It is thought that black holes store and distribute graviton particles. The source of which comes from a consideration from Hawking radiation; as a particle-antiparticle pair splits at the event horizon, antiparticles (associated with antimatter) remain inside. With gravitinos being a supersymmetric particle that can (questionably and arguably) assume the role of antimatter, they too reside inside, but not for long. Considering the first direction of composite theory, as two gravitinos collide and become a graviton and a photon, more and more gravitons would be made inside a black hole. With gravitons being bosons under Bose-Einstein distribution, they could either be disbursed like a gas in a box or lounging around somewhere inside.

Black hole well theory is ingenious as it is ambiguous. Because black holes are "black," observation of the inside leaves us with probabilistic outcomes and shrugging shoulders. But on the optimistic philosophical note, this may even consider black holes as central nodes of interdimensional transportation in brane theory.

Take the black hole completely away, only having a coupled chunk of gravitons, their self-gravitation would simulate a quantum black hole, which would only increase with more and more compiled gravitons. This is similar to the creation of Kugelblitz black holes (black holes purely made of and by radiation). It even considers the wave-particle duality of gravity waves, that gravity waves are made by unstable black hole binaries.

GRAVITATIONAL WAVE-PARTICLE DUALITY

This next chapter is on the novel theory of gravitational wave-particle duality: the duality between the quantum graviton and the astronomical gravitational wave. Because gravitons are only acknowledged to have a wave-particle duality with gravitational waves, but yet not proven to have so, this chapter is rather investigative and tentative than a widely accepted and conventional theory.

This theory of the graviton attempts to conceptually unite string theory and loop quantum gravity, to say that gravitons are compatible in both theories. Although they are born from string theory, the acknowledgement of gravitons in LQG only extends to behaving like neuron signals between the spin networks in the lattice. But how can a graviton, in both theories as part of the spacetime brane, can behave as a kinetic particle in a propagating wave?

This theory considers two of the current theories of gravitons: brane theory and black hole well theory; and how it ties in with LQG instead of string theory. Since string theory can describe the graviton as a field particle through gravitational interaction, it would be beneficial to depict the graviton as a field particle in the context of the quantum spacetime geometry.

In LQG, canonical spacetime (canonical meaning fundamental) is described by the Wheeler-DeWitt equation, which can be expressed as the sum of overlapping quantum curvatures within a single entity of spacetime. Since canonical spacetime represents *flat* spacetime, a gravitational wave is a ripple, a *perturbation*, upon the fabric. Much like the ocean, the waves occur at the upper surfaces of the body of water, while the deepest layers remain unaffected.

This applies graviton brane theory with the quantum layering of spacetime via the ADM Formulism, the bottommost layer being a graviton particle field. The embedded gravitons, given that they exceed their "fission energy" either by randomness or by field agitation, "leak" themselves out of the particle field and into the uppermost energy layer. This gives these freed gravitons on the top layer a "flat stage" approximation: assuming they do not curve spacetime, and that they are merely actors upon a rigid stage.

Therefore, the Hamiltonian of the uppermost spacetime layer represents a gravitational wave / a moving graviton:

$$\frac{\hbar^2}{m_g} g^{\mu\nu} \partial_\mu \partial_\nu \psi = \hat{H}\psi,$$

which preserves the Wheeler-DeWitt equation, $\hat{H}\psi = 0$, for the quantum spin projection $m_s = 0$.

This would mean that, with a zero-spin projection value, the graviton upon flat spacetime has a "cloak of invisibility:" despite being able to have zero spin and still be present in the space, a zero Hamiltonian suspects the graviton is not even there. Well, in physics, either the thing is there or not.

With gravitons having 2 degrees of freedom, its constraint is a zero-spin projection: a graviton cannot have zero spin to have any physics at all. Therefore, the Hamiltonian of a wave graviton is spin-dependent:

$$\hbar c \sqrt{2|m_s|}\, g^{\mu\nu} \partial_\mu \partial_\nu \psi \sim \left(m_g + i\partial_t\right)\psi,$$

where the left-hand side resembles the string length of a particle, in this case, a wave graviton.

This equation considers the perturbed geometric warping upon spacetime in form of a gravitational wave as a propagating graviton. Therefore, it is arguably the coup de grace of gravitational wave-particle duality. However, for gravitational waves from a black hole merger, the graviton wave equation also asks the question: *where do these wave gravitons come from in this case?*

117

Since the astronomical gravitational waves come from the collapse of an unstable black hole binary, an emission of a gravitational wave graviton would take place at the moment of black hole merger: when the two collapsing black holes morph into one, sending the final seismic spike in the spacetime fabric.

This applies the black hole well theory into this wave-particle duality: since gravitational waves come from the "chirp mass:" a specific, singular mass merged from the two original black holes that sends the final seismic spike, a graviton must leak from the chirp mass black hole.

Should a black hole behave as a potential well containing gravitons, the size of the black hole increases the maximum amount of gravitons it can store, increasing the number of "graviton states" (each particle state represents the presence of a particle; each particle has a state). As bosons, the individual particles initially in their own states are free to couple together or to condensate down into a singular common state. But during black hole merger, the particles may be jostled enough to be "shaken" back to their individual states.

What the gravitons may be precisely doing inside the black hole at the moment of merger is of no concern here, but the particle at the uppermost state (at the event horizon) would leak out, thus comparing the quantum energy of the emitted graviton to Hawking radiation!

As a particle, the resulting wave function from the equation is spatially symmetric, meaning the function looks the same for infinite forward and backward positions in a single time frame. Given a propagating ripple from a collapsed black hole binary, the arrow of time points forward with the propagation. When you flip through the time frames like a deck of cards, the wave is at a different spot every time. This makes the wave function time *asymmetric*. Because both space and time are rippled by the gravitational wave, time shall also be symmetric in the function. But how can we instill time symmetry with an irreversible arrow of time?

Philosophically, while we live our lives with time moving forward irreversibly, recollecting memories is a form of regressive time travel, as it is also imaginary. As we "recollect" past moments of the wave's propagation, we would have to use "negative imaginary time:" a mathematical trick where time is treated as another dimension of space. In this case, "negative space," or infinite backward positions. This way, the gravitational wave can be described as a moving particle in the quantum sense with the following function:

119

$$\psi \sim \varphi e^{-\varphi^2},$$

Where φ is a dimensionless "spacetime parameter" that considers the simultaneous progression of position and time.

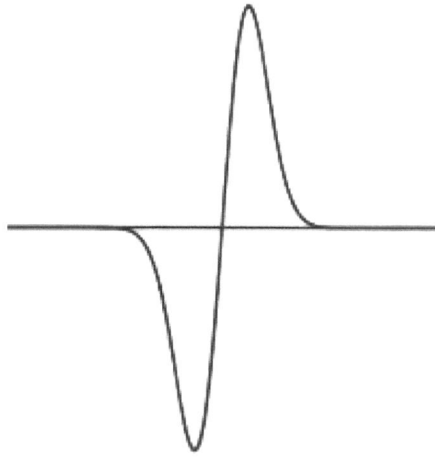

A propagating wave graviton (purple) overtop the spacetime equilibrium (black).

THE FUTURE

If I had my own definition of science, it would be this:

Science is the perpetual investigation of a better logical description of the universe and how nature works, which must be refutable in order to achieve its trustworthiness obtained by human observation and understanding.

As a man of science, of faith and philosophy, I stand by my definition in search of a concrete, sensible idea of the mechanics of nature. While many scientists, such as my colleagues and professors from my studies, would want to drive forwards on this highway of theoretical understanding, one might want to go on a scenic route, detour or even a back-track to make sure no ancient or alternative knowledge is left behind or unseen.

One just might appreciate religion without the holier-than-thou dogma, be in awe and enlightened by the ancient and (at first glance) primitive worldviews, and thus "connect the dots" between past knowledge and the present-day questions in theoretical physics. Perhaps a "neo-Renaissance Age" might come to pass, and should today's scientists stick with the one-way mainstream convention, they might find themselves in a traffic jam, cursed to be blocked in endless blockage.

To continue this quest to redefine gravitational theory, one would have to understand how the philosophers and scientists from long ago viewed gravity and the cosmos, and not dismiss it prematurely. If it were not for the scientific revolutionaries referring back to the ancient texts and philosophies, we would not have known that everyday free-fall was surface gravitation. How one simple action: looking up at the starry night sky, watching an apple fall, or seeing how a fire burns the kindling down to ashes; leads to revolutionary theories of constant free-fall, bendable space and time, and evaporating black holes.

The future is much of an illusion as the concept of time. But while hindsight is always 20/20, the future is as blind as a bat. We cannot know for sure whether the future of gravitational theory or scientific understanding would continue on their current trajectories. Maybe that is why people still cling to religion, because the future is certain to those who believe (there is an afterlife or a second life). Not everyone can accept the uncertain nature of the universe and this everchanging status of modern science. But that is why science is trustworthy to begin with; human intelligence has modified scientific theory in order to meet the technological standards of the present day. It is trustworthy because *we* can conduct scientific research.

As for me – a man of science, faith and philosophy – I do not let my scientific knowledge quantify my beliefs. And furthermore, I do not let my beliefs dogmatize my knowledge of the universe. Much like church and state, there is a sure difference between believing and knowing: I do not know that God exists, but I believe it. I don't have to believe in evolution because I know it to be scientifically true. When approaching a tentative theory such as quantum gravity, there is no reason to use the word "believe." If provided a trustworthy theorem or mathematical representation, or a logical reasoning that refers to a past theory, there is no believing, but rather knowing – or trusting.

However, the problem quantum gravity is facing is its progressive lack of "trustworthiness." String theory and its 11 dimensions work for the Planck scale, which is different from our macroscopic scale. How can we trust in string theory if we cannot measure what the math produces? Loop quantum gravity considers spacetime as a foaming lattice of spin networks: these grainy quanta of spacetime. How can we trust in that theory if we cannot harness, or pluck, a single spacetime quanta like we can with electrons?

Sometimes, the concept of knowing has to convert itself from physical proof to mathematical proof. Just like the Higgs boson, a theory with the substantial mathematical proof will in due time be given physical proof, given the right technology and the right approach. On the other hand, much like a basic concept of free falling or the simple action at looking up at the starry night, physical proof will in due time be given mathematical proof, given the right technology, the right approach, and the ability to take the road less traveled.

One way or another, you will get to where you are going.

ABOUT THE AUTHOR

Born in Natrona Heights, Pennsylvania (a suburb of Pittsburgh) in 1997, Noah Matthew MacKay was raised in eastern North Carolina since 2000. In 2005 he was diagnosed with Asperger's Syndrome at the age of eight.

MacKay started a hobby of writing fiction novels in 2013 (at the age of fifteen) and keeps up with it ever since. As a novelist, MacKay wrote five books from his fantasy series titled *Age of War* between 2016 and 2020. Since 2015, Noah MacKay also writes poetry in the

German language. His noteworthy poetry publications are in the following collections: a "best-of" collection of his 2015-2019 poetry and autobiographical account *Einfach ich* (Simply Me) and *Monster* (Monster), both in 2020; and the latest collection *Asyl* (Asylum) in 2021.

In addition to writing novels and poetry, MacKay wrote four volumes of the mathematical physics series *The Theory of Physics* in 2020, each volume discusses respectively on classical mechanics, electromagnetism, select topics of modern physics and quantum mechanics. In 2021, MacKay wrote *Quantum Particles of Gravity*: a layman's course in gravity theory from the ancient philosophies to the hypothesis of gravitons; along with a direct German translation of which and a more mathematical rigorous version titled *The Theory of Gravity*. Alongside his physics books, MacKay wrote the English and German versions of the book *Germany in the Twentieth Century*, which recounts the national phases of 20th century Germany and how select literature and cinema conveyed the tenor of the times.

In 2020, Noah M. MacKay graduated magnum cum laude from East Carolina University. He acquired a Bachelor of Science degree in physics, along with a Bachelor of Arts in German language and literature. As an undergraduate, he co-founded the university's first official astronomy club, became a member of the Sigma Pi Sigma and Delta Phi Alpha honor societies, and co-authored a published article on Weimar Republic-era poet and satirist Erich Mühsam.

MacKay is currently studying physics at East Carolina University for a Master of Science, as he is preparing for his thesis work on quark-gluon plasma. In the future he plans to obtain a doctorate and become a published theoretical physicist.

www.ingramcontent.com/pod-product-compliance
Lightning Source LLC
Chambersburg PA
CBHW041313180526
45172CB00004B/1080